THE LAWNCHAIR ASTRONOMER

THE LAWNCHAIR ASTRONOMER

Gerry Descoteaux

A Dell Trade Paperback

Book produced by **David Kaestle, Inc.**
Art Direction: **David Kaestle, Richard DeMonico**
Chapter openers and page head graphics: **Arlene Gertzulin**

A DELL TRADE PAPERBACK

Published by
Dell Publishing
a division of
Bantam Doubleday Dell Publishing Group, Inc.
1540 Broadway
New York, New York 10036

Library of Congress Cataloging in Publication Data
Descoteaux, Gerry,
        The lawnchair astronomer/Gerry Descoteaux.
            p. cm.
        Includes bibliographical references.
        ISBN: 0-440-50696-4
        1. Astronomy—Popular works. I. Title.
QB44.2.D48   1995
522—dc20   94-39694
                                          CIP

Printed in the United States of America
Published simultaneously in Canada
June 1995

10 9 8 7 6 5 4 3 2 1

Contents

Acknowledgments

Thanks to my wife, Lynn, Dennis Brisson, David Brooks, Kelly Prothero, John Brennan, Edward Vesneske Jr., Ling Lucas, Eric Wybenga, Richard Lovinson, the U.S. Naval Observatory, and NASA.

Dedication

To my daughter Jillian

With all my love and hope
May your skies always be filled
With the wonderfully bright jewels
That will always serve
To remind me of you.

Just a Bunch of Stars?

There are thousands of professional astronomers who work at the National Aeronautics and Space Administration as project designers and managers, and in universities as professors, teachers, and research scientists. But this handful of men and women are just the tip of the astronomy community: as a matter of fact, the number of amateur astronomers substantially dwarfs that of professionals. What bonds amateur and professional together is their shared underlying awe of the universe and everything in it.

Many of the greatest astronomical discoveries have been made by amateurs. Working together with their paid counterparts, they have contributed a great deal to our overall knowledge of the universe. Therefore, I am honored to be recognized as a member of this group. Not everyone can devote a career to the pursuit, but anyone can partake in this lifetime of wonders and just plain old-fashioned fun. Anyone can be a Lawnchair Astronomer!

What's so exciting about looking at a bunch of stars? Though the answer may not be obvious to the uninitiated, to anyone who has tasted the thrill of seeing a new galaxy or star cluster for the first time, it's quite simple — as simple as going out into the backyard and looking, really looking, at what's up there. Our ancestors were awed by what they saw in the night sky. That sky hasn't changed significantly in the relatively short period — cosmologically speaking — since their time, and neither has the sense of wonder that comes from seeing the tapestry of stars and wisps of light that rain down on our backyards each night.

Stars come in a wide variety of sizes and colors, along with a host of other attributes that may not be noticeable in a quick glance around the evening sky. Many star systems consist of multiple suns, orbiting around each other or around a common center of gravity. Binary and triple star systems are quite common in our galaxy as well as in others.

Star clusters — great, often tightly packed conglomerations of stars — are another easily viewed phenomenon visible to the Lawnchair Astronomer. Some of the most famous, like the Beehive or the double cluster in Perseus, are the stuff that telescope ads are made out of.

The great nebulae, like the one in the Orion constellation, are on every Lawnchair Astronomer's must-see list. The thrill of seeing a great nebula for the first time is truly indescribable. It's one of those things in life that you have to experience for yourself.

Besides the star clusters and the thousands of wispy nebulous objects, other amateur stargazing targets include the eight planets, hundreds of asteroids, and several recurring comets that are occasionally available for viewing from your backyard. In addition to these — saving the best for last — there are the external galaxies.

Just as in our own stellar neighborhood, the external galaxies are home to all of the different types of heavenly sights I've already mentioned. Today's Lawnchair Astronomers, armed with the best equipment, are privy to worlds that a mere hundred years ago were hardly dreamed of — and there are perhaps as many galaxies left to explore in the universe as there are stars in the Milky Way!

So you see, there are plenty of sights to thrill you, too, in the dark, star-filled skies of our world. On a clear, moonless night, the stars you can see from a dark-sky location will number in the thousands. With a pair of binoculars, their number increases a hundredfold. With a small to medium-size telescope, you'll gain access to nearly a *trillion* stars!

Yes, there's much more up there than "just a bunch of stars." There is so much to see and learn, it could easily take several lifetimes to experience everything astronomy has to offer. So try it out for yourself. Turn off the television and get outside this evening, and look — really look — at the sky. Bring along friends and see what develops. Remember to dress warmly, and most of all — enjoy yourself. Happy stargazing!

LAWNCHAIR ASTRONOMY

Best When Shared

Look at it this way: much of what can be seen in a very dark sky is completely accessible without the use of a telescope. As a matter of fact, some of the most wonderful astronomical targets and events are best seen with your eyes alone. Meteor showers, eclipses, bright comets, and the star pictures we call constellations are but a few of the wonders you can see in a dark sky from your own backyard!

One of the most enjoyable experiences I've ever had while stargazing occurred a few summers ago as I was sitting in a large country field. A friend and I, accompanied by nothing more than a pair of lawnchairs and some cool drinks, sat for several hours shooting the breeze, so to speak, while reviewing the constellations we were each familiar with. Luck being on our side, we encountered a fairly active meteor shower that evening. We stayed until clouds rolled in around two in the morning. To this day, every time we get together, we warmly reminisce about that wonderful night in August, out under the stars.

Being out there, under a truly dark and brilliantly jeweled sky, is in itself almost a religious experience. You feel this as soon as you get out there, especially on a moonless, cloudless midnight. The tranquility of the night is often very striking compared to the hustle and bustle of daylight hours. But you feel the adrenaline rush as you come to realize that you're not alone in your enjoyment of the night's solitude. There is a whole community of nocturnal creatures that share your delight in the starlight and shadows.

If you want to try something completely different from the regular late-night, channel-surfing, insomniac boogie, spread a sleeping bag out across the front lawn or out on a distant prairie.

Lay back and really look at the heavens. See the myriad colors of bright and dim stars. See the blurry little areas that with an inexpensive pair of binoculars or small telescope resolve into thousands of pinpoint sapphires. It is something you must experience for yourself. And that's all the convincing you'll need to count yourself part of the club — of Lawnchair Astronomers.

GREAT MOMENTS IN LAWNCHAIR ASTRONOMY

Stonehenge

Evidence of ancient astronomy dates from between 3000 and 2500 B.C.: the great stone circles and monoliths found throughout northwestern Europe, the most notable, of course, being Stonehenge in England. Nomad Druid priest-astronomers marked the daily paths of the Sun, Moon, and stars throughout the year in the construction of this and other ancient observatories. It is thought that by 2500 B.C. Stonehenge was used to predict eclipses. It wouldn't be until A.D. 1000 (more than 3000 years later) that similar activities began in the New World, where the Lawnchair was soon to gain a permanent foothold in society.

THE TOP TEN NAKED-EYE LAWNCHAIR TARGETS

Information on when and where to look for many of the universe's best naked-eye sights can be found in the monthly astronomy-based journals like *Sky & Telescope* and *Astronomy* magazine, where skycharts and news stories will direct you to the month's best sights. Below is a list of the ten most interesting naked-eye Lawnchair-accessible sights.

1. **Meteor Showers** Easiest to enjoy and require nothing more than a sturdy lawnchair and a clear, moonless sky.

2. **Auroras** Northern skies are where you will thrill to this sight, as often green-tinted curtains of light reach down to the horizon.

3. **The Milky Way** Another job for your most comfortable lawnchair, best late on a warm summer night in the darkest sky you can possibly lay under.

4. **Lunar Eclipse** The only time I'd ever recommend a campfire while out Lawnchairing, this long-duration event is best enjoyed with large gatherings of fellow Lawnchair worshippers.

5. **Mercury** This one is a challenge and requires a clear open view to either the eastern or western horizon — just before sunrise or just after sunset, depending on where Mercury is in its orbit around the Sun. Seeing Mercury with eyes alone has long been considered one of the most intriguing Lawnchair trophies.

6. **Mars** Mars obviously stands out because of its famous red tint, and depending on where it is in its path around the Sun, will mean either breakfast or dinner out on the lawnchair.

7. **Venus** Often associated with increased UFO reports, especially when it follows the Sun in the evenings, Lawnchair enthusiasts can take comfort in knowing that it is just Venus, the brightest of all the planets.

8. **Saturn** From a comfortable lawnchair its warm, often yellowish color is easily apparent, and it shines as bright as the brightest stars.

9. **Jupiter** Bright, or rather brilliant, this beach ball of a planet stands out because of its sheer magnitude, the brightest object other than the Moon that you'll ever see in a midnight sky (because you'll never see Venus at midnight).

10. **Satellites** Too slow to be meteors and too high to be jet airplanes, these man-made orbiters are lit the same way the Moon and planets are — by the Sun — and can be seen crossing overhead nightly under clear, Lawnchair skies.

Eclipse
(or Just Another Excuse for a Party)

Eclipses have a way of encapsulating special life moments. Just as many people, myself included, remember exactly where they were when John F. Kennedy was assassinated, the same holds true for those lucky enough to witness an eclipse. One of nature's best excuses for a party, all you need to enjoy one of these spectacular celestial events is your favorite lawnchair and a dozen of your best buddies. Even without the promise of snacks, drinks, and half a dozen pepperoni pizzas, these parties are usually well attended.

The mechanics of an eclipse are easy to understand. In the case of a lunar eclipse, the Earth finds its way between the Sun and Moon, and projects its shadow against the Moon. In the case of a solar eclipse, the Moon physically passes in front of the Sun, thus blocking it from our view. In ancient times this was not always understood. For example, when Columbus, during one of his many adventures in the New World, had been captured by a tribe of natives, he used his knowledge of astronomy to free himself. As a navigator he was privy to astronomical charts and statistics, and was aware of an impending solar eclipse. Thinking quickly, he pretended to be a deity and convinced the natives to let him go once he produced the darkening of the Sun, which they worshipped as the living manifestation of their god. With their limited experience of astronomy and their belief of such things as angry gods, Columbus did indeed seem to command the heavens. Today we understand everything about eclipses and are able to predict them with incredible accuracy.

How to View an Eclipse Without Going Blind

Staring at the Moon during a lunar eclipse is perfectly safe. Precautions must be taken, though, when attempting to watch a solar eclipse. During most of a total solar eclipse (when the Moon completely covers the Sun) or a partial eclipse (when the Sun never quite gets covered) or in the rare case of an annular eclipse (like the one we experienced here in North America in May 1994, when the Moon's disk at totality fit within the disk of the Sun, thereby leaving an outer ring of sunlight exposed) you should never look directly at the Sun. Even the smallest exposed sliver of sunlight can do irreparable damage to your eyes, and cause, in the worst case, blindness.

The only time that you can look directly at a solar eclipse is during the totality stage of a total eclipse, which completely covers the Sun's disk. And at most, you may look only during the few minutes of totality before the Sun reemerges from behind the Moon. At any other time during a solar eclipse, you must use one of the recommended methods and/or devices especially designed for watching eclipses.

The most popular viewing accessory today seems to be the specially designed Mylar eclipse glasses, which are generally abundant the day after an eclipse has occurred. Constructed of an aluminized Mylar material set within throw-away cardboard frames, these can be purchased for as little as $1.50 a pair.

This same Mylar material is also used in professional-grade solar filters designed to cover the front end of your binoculars or telescope. This accessory allows for more professional observations of solar eclipses, as well as for daily solar observation during which sunspots are easily detectable by Lawnchair-grade instruments. Also, unlike the #14 welder's glass that is often recommended for this purpose, Mylar filters perform much better optically.

Finally, an easily constructed device is the projection plate or box. Pierce the center of a paper plate with a needle or pin, then place it nearly at arm's length in front of you, angled toward the eclipsing Sun at your back. From there, project the Sun's image through the hole onto either another plate below or right down onto the ground. This is the very safest way to watch a solar eclipse.

Plato's Spheres

Greatly influenced by Plato's (428–347 B.C.) metaphysical concept of the perfection of circular motion, the Greek Lawnchair set tried in vain to explain the motions of objects they believed were governed by divine designation. One of their first theories employed geometric models which used spheres and circles to predict what they were seeing from their Lawnchairs.

Eudoxus of Cnidus (408–355 B.C.), a Platonian scholar (and closet Lawnchair Astronomer), believed that the planets were attached to concentric rings or spheres centered about the Earth. What his model failed to predict were the changes in brightness of the planets. Even so, his model was incorporated into Aristotle's cosmology, during the 4th century B.C.), which attempted to describe the universe as being centered around the Earth — a geocentric view.

See It First and You Get to Name It

Magellan, Skellerjup, West, Halley, Tempell, and Tuttle are not the names of the newest punk, rap, and heavy metal bands on the MTV heavy-rotation list. They have a more lasting fame: they're just a few of the thousands of people who have discovered comets and other celestial objects, and named them, usually, after themselves! One of them, Charles Messier (1730–1817), has a whole catalog of stellar jewels named after him.

In eighteenth-century France, Messier hunted comets — a very meticulous line of work that requires in-depth study of the sky. A problem he often encountered, though, was confusing cometary targets for the many fuzzy, ghostlike nebulae and thick star clusters that pepper the night sky. To alleviate this pain in the eyepiece, he charted and cataloged these many mysterious objects, which today are known by astronomers everywhere as the Messier objects or the Messier catalog.

As a Lawnchair Astronomer, you will need to use this lingo if you're to fit in. These 110 objects are referred to by their Messier numbers. For example, M-31: the Great Cloud, or Andromeda Galaxy. This is one of the prime Lawnchair targets. At 2.2 million light-years away, it is the most distant object visible to the naked eye. With both binoculars and small telescopes, this fuzzy patch of light resolves into a bigger fuzzy patch, the sister galaxy of the Milky Way.

In Messier's day, telescopes were still a relative novelty. They could hardly compare in power to even the low-priced amateur instruments of today — which makes the Messier catalog an excellent place for a Lawnchair Astronomer to begin the hunt for astronomical treasures.

And who knows, you might even find a few that Messier overlooked! Okay, so he was there before you, but there are still plenty of things left out there for you to find and name. Dozens have been discovered in the last twenty years. Like everything else, it's a matter of being in the right place at the right time.

The Babylonians

Some of the first great reaches in astronomy occurred in ancient Babylonia. Between 1800 and 400 B.C., the Babylonians developed a fairly accurate calendar based on the motions of the Sun and the phases of the Moon. We know this from deciphering ancient cuneiform tablets, which use a sort of picture-language. Archaeologists have tentatively identified sketches on these tablets of the earliest Lawnchairs, usually seen being carried by four scantily clad servants.

These graphic texts also show that these early scientists achieved an accuracy of astronomical prediction to within a few minutes of perfection. In the following four hundred years they concentrated on increasing this accuracy by predicting the exact time that the first crescent moon could be seen — also a popular and longtime Lawnchair Event. The Babylonians compiled extensive numerical data and observational tables which allowed them to eventually predict lunar and solar eclipses with a great degree of precision.

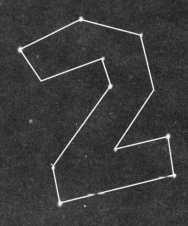

TOYS OF THE TRADE

THE LIST
· · · · · · · · · · · · · · · · ·

Remembering to bring along everything you will need is, to say the least, the best way to start — especially if your observation point is not exactly local! You'd be surprised how often crucial things get left behind. The roads to and from my favorite sites have been privy to some very inspired cursing, sometimes in multiple dialects and languages, as I suddenly realize I've forgotten to bring along one thing or another. That's why the list of my Lawnchair toys is prominently taped to my telescope!

I have a friend whose list begins with plenty of his own special brand of thirst-quenching fluids. He also has on his list lawnchairs (of course!), blankets…and the ice chest. (We can't have those thirst-quenching fluids getting warm now, can we?) Then there's:

★ **Insect repellent.** Of all the things that might find their way onto your list — from a first-aid kit, to a name tag in case they find the body, to mass quantities of everything from pizza to burritos — this is the item most often overlooked. But it shouldn't take more than a few summer nights outdoors to convince you that this is, after all, the insects' domain, not yours. If you don't bring along an adequate "force field," you are dinner! In big, bold letters, bug spray is quite prominent on my list. Save yourself the learning curve and add it to yours. You'll thank me in the morning.

★ **The right clothes.** Remember, even on a hot summer day, the temperature can drop quickly and dramatically by nightfall, so you'll want to dress accordingly.

Items for the posteighties yuppie set might include a Humvee or other 4 by 4 carriage for you and your friends to get your toys atop that far-off peak. A cellular phone probably wouldn't hurt either, should your off-road exploits leave you in a compromising situation. More astronomically pertinent equipment, on the other hand, will include:

★ **A star map.** These come in a variety of styles and formats. A copy of any recent *Astronomy* or *Sky & Telescope* magazine will point you in the right direction. As a matter of fact, each of these magazines publishes a monthly star map in their centerfold slot. Included in the Lawnchair Astronomer is just such a map. Use it to find your soon-to-be-favorite celestial treasures!

★ **A red-tinted flashlight** in order to look at the map without ruining your "dark-adapted" eyesight. Just take a regular flashlight and mask it with red cellophane. (One of the things you'll learn to do is to avoid any source of light. The more light-sensitive your eyes become, the better you will see those wonderful, though often dim, splashes of light floating across your eyepiece.)

Whatever you think will add enjoyment to your evening out under the stars should be placed on your list. There are no bad choices. So my advice is to make a list, and double-check it every time you go out to stargaze. Do this, and your armchair curiosity will turn into guaranteed Lawnchair success.

My, What Big Eyes You Have: Binoculars

Looking across the night sky with your eyes alone, you're sure to see plenty of stars and other celestial inhabitants. The next best thing, though, is to add some type of optical aid to the equation. Without spending incredible amounts of money, a pair of quality binoculars can increase your enjoyment of the sky tenfold and reveal an even more fantastic universe for you to peek in on.

A pair of binoculars can open your eyes, for instance, to an enormous Moon, with craters and mountain ranges that would be dramatically discernable. Then there are the bright planets, including Jupiter and its largest moons, as well as Saturn and its ring system. In addition to these relatively easy targets, binoculars can reveal countless intriguing patches and streaks of ethereal light that in long-exposure photographs reveal themselves to be external galaxies or exotic gas and dust remnants of deceased stars.

The main consideration when purchasing binoculars is their capacity to reveal dim objects. To do this, they must be able to gather light efficiently. The technical factor that limits this ability is *aperture*, or the size of the front lens. For astronomical purposes, the rule is the larger the aperture the better.

At a relatively modest price range ($80 – $250), 50mm field glasses are the most appropriate. Camera shops, department stores, and magazine ads for mail-order houses are a few places you can shop for a decent pair. The key word here, though, is shop — as in shop around!

Typically they'll be listed as 7 x 50 or 10 x 50. The first number refers to the magnification power of the binoculars. The 7 x 50 therefore, has an aperture of 50 millimeters and magnifies an object by seven. There are more expensive, higher-powered binoculars, but the 7 x 50's or 10 x 50's are quite adequate.

Another binocular component that correlates to optical quality is the complex set of prisms, which bend and fold the light path in order to deliver a right-side-up image to the eyepieces. There are two types: roof prisms, which use a straight tube design, and Porro prisms, which incorporate an N shape or zigzag design.

For considerations of function, quality, and astronomy, Porro-prism binoculars are generally preferred over roof-prism models. You must be careful, though, since lower-cost (cheap) Porro-prism binoculars can sometimes be disappointing, while the medium- to higher-priced models will do the job nicely.

Another feature to consider is the type of coating that is on the lenses of your binoculars. Most binoculars have at least a single coating on the exterior lenses. (This coating is visible as a bluish tint; it's not there to protect the lenses from scratches, but to increase their light gathering capabilities.) Some manufacturers coat both sides of every lens, while others coat just the external surfaces. Higher cost and quality usually mean more extensive coating of the lenses and prisms — which directly correlates with how well the binoculars will perform as an astronomical instrument.

Coating offers no physical protection for the binoculars, so be sure to buy the accompanying travel case. It's one item you shouldn't skimp on, since it will serve to protect your purchase for years.

A tripod for mounting the binoculars will also help you to avoid that morning-after stiff neck syndrome. Primarily, though, it will greatly increase your chances of accurately focusing on and observing your target. Tripods come in a variety of styles and sizes. Desktop models work well from a deck or picnic table. In any case, a good tripod should hold your instrument firmly and solidly, even in a brisk wind.

THROUGH THE LOOKING GLASS

Every August, usually around the middle of the month, a small town in Vermont is descended upon by hordes of men, women, and children. In tents, campers, and trailers, these folks arrive from around the world to partake in what has become the best-ever star party in northern New England. In case you're not familiar with this event, it is called Stellafane, and it is put on by the Springfield Amateur Telescope Makers Club of Vermont. Each year the crowd has grown bigger and better, and organizers would want you to know that this year's event will once again surpass previous star parties on Breezy Hill. Thousands of amateur and professional astronomers and telescope makers converge on Springfield each year to play with...er, field test and display their homemade and store-bought instruments.

Stellafane is but one of several dozen annual amateur and professional astro get-togethers across the country. The majority of these events are open to the public, and in the case of those like Stellafane, can involve camping out over the course of a weekend or, in some cases, as long as a week or so. They are usually well advertised in the astronomy magazines, where you can also get reservation information and schedules.

More than Pretty Colors and Shiny Buttons

Seldom seen in today's whiz-bang radio and television advertisements are big-time promos for telescopes. As a matter of fact, there has never been a giant media blitz advocating that you run right out and buy your very own "super-sensitive, dark-adaptive, automatic, slicing, dicing, combination deep-sky telescope, can-opener and fish-stripper, which will fit nicely behind the seat of your Chevy pickup!" It's a shame, too, 'cause I bet they'd sell a million of them.

There are several sources for astronomy-related equipment, though, and since there is also a range of prices you'd be well advised to research your purchase thoroughly before laying out your cash. You'll also want to determine what you intend to do with your telescope or binoculars. If you've never owned or used a telescope, you may find the vast selection confusing.

Basically there are three kinds of telescopes: refractors; reflectors; and the combinations, or catadioptrics. Each comes in a multitude of styles, types, and sizes, in a wide price range. A good beginning telescope, though, is the standard one- to three-inch refractor. Refractors use lenses to deliver an optical image to your eyes, while reflectors use mirrors, and the combination type uses both lenses and mirrors. Each has their advantages and disadvantages.

Before you pick out a telescope, though, you must think carefully about where your heart lies. Will the many bright stars, planets, and the brightest of the deep-sky objects that can be seen with a refractor be enough to satiate your interest, or are you really interested in the dim little fuzzies that only a larger, reflecting telescope will allow you to see?

The Refracting Telescope

A refracting telescope is that classic, ship captain's spyglass. It consists of a closed tube with a large object glass at one end, often called the objective lens, which is sometimes composed of two or more separate components. Light from a star or planet passing through the telescope is refracted by the objective lens to a point of focus at the lower end of the tube. Here, a wide variety of eyepieces — for example wide-angle or close-up — can be used like magnifying glasses to enlarge the incoming image. With a refracting telescope you can see all the bright planets, the rings of Saturn, the cloud belts of Jupiter — as well as its large moons, which change position nightly. You can stare for hours at some of the bright stars, each with its own strikingly vivid color. The trillion pinpoints of the Milky Way are also visible, as well as the famous star clusters and bright external galaxies. The quality refractor will easily be able to detect those special Lawnchair favorites. In a dark sky, the bright, smoky, nebulous objects, such as Orion's nebula and the great Andromeda Galaxy are terrific, even with the most inexpensive, small refractors.

You'll find these simple telescopes in some department stores at around fifty to two hundred dollars. Also, local camera shops often have telescopes and binoculars in stock. As opposed to the toy-quality instruments found in many discount stores, here and in dedicated telescope shops you can expect to find brand-name, semiprofessional products. Today, Meade and Celestron telescopes are quite popular and are highly rated by both amateur and professional astronomers.

As a novice, your best move is to pick up a copy of *Astronomy* or *Sky & Telescope* magazine at your local newsstand. Here you can easily get a feel for quality versus price on a wide variety of instruments. You can shop through the mail or by phone, and even by computer through any of the major computer networks, even if it's just to use those prices to compare with the one for that shiny telescope that's caught your eye in the local department store window.

If your interest lies in just a cursory viewing of the planets and the brightest deep-sky objects, then you're best advised to stay with the "under a few hundred dollars" range of telescopes. If, however, you'd like to delve deeper into astronomy and the universe, and see not the clusters of stars but the dim, firefly groups of galaxies, perhaps even photograph some of them or the dim asteroids, or even the furthest planets, then you might consider owning one of dozens of top-quality professional-grade telescopes that are available today.

You could also build your own instrument if you are so inclined. That by itself could take up a whole book. As a matter of fact, you can find several telescope building books, containing all the directions you need, at your local library. Pick up one of the latest telescope makers' handbooks or even one of several buyer's guides. The best decision you can make will hinge on the knowledge you have, before you buy, about the instrument you wish to build or purchase.

The Reflecting Telescope

A reflecting telescope focuses light rays by way of a large, curved "concave" mirror at the bottom of a tube that is usually larger than you would find in a refracting telescope design. Light striking this mirror is reflected back up the tube to a secondary mirror that is flat but mounted at a 45-degree angle, which again deflects the converging light rays 90 degrees into the eyepiece. Named after its inventor, Isaac Newton, the Newtonian reflecting telescope revolutionized astronomy.

With a large reflecting telescope, not only are Saturn's rings visible, but so are the major divisions in the ring system. In a quality reflector you can study the cloud belts of Jupiter, and watch, over a period of a few hours, as the Great Red Spot — a giant, Earth-size storm — traverses the globe right in front of your eyes. To see those distant galaxies and million-member star clusters, or the faint supernova remnants and other compelling sights, you'll need a large-aperture reflecting telescope.

Those of you who are critically afflicted, and out for the "aperture cure" available from the larger and usually costlier reflecting telescopes would be best advised to check out the magazine advertisements first.

A decent quality, though very basic, reflecting telescope may cost as much as fifty to sixty dollars per inch of aperture, but in a top quality, computerized, worm geared, star tracking, roto-rooting, Winnebago-type instrument, this price could quadruple. You could easily find your way into buying a telescope that would almost seem to require a technical degree in electronics to operate it, never mind a price range and payment plans that could compete with your mortgage. If you're looking for a lot of bang for your buck, you'll need to compare your desires to your wallet and look for the best deal.

The light bucket — often the most envied and largest of the backyard telescopes — is most commonly found in the eight- to seventeen-inch

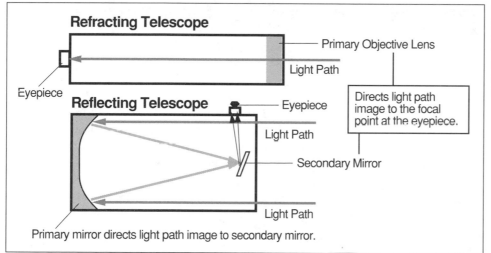

Refracting Telescope

Primary Objective Lens

Light Path

Eyepiece

Reflecting Telescope

Eyepiece

Light Path

Secondary Mirror

Directs light path image to the focal point at the eyepiece.

Light Path

Primary mirror directs light path image to secondary mirror.

The two basic telescope designs use lenses to magnify and focus images that arrive via either an objective lens or a combination of mirrors.

apertures. The choice of deep-sky addicts, they provide the greatest light-gathering power in Lawnchair-grade instruments. The down side is that these large tubes often require you or one of your best friends to own a truck. Priced from several hundred dollars to as much as ten thousand dollars or more, these large telescopes will show you the universe's most highly guarded secrets.

Consider all these factors before running out and buying any astronomical instrument. Whether a small, toy-grade refractor or a twelve-inch, fork-mounted slicer-dicer, make sure you get the one that will actually let you look at the stars!

Catadioptrics/Schmidt-Cassegrain Telescopes

The large, fancy, high tech, often computerized, Schmidt-Cassegrain catadiotropic telescopes are at the top of the scale in Lawnchair-grade telescopes. But I strongly suggest that you know what you're doing before acquiring one of these. These expensive instruments, often seen adorning the lawns of wealthier enthusiasts, require that a knowledgeable butler be close at hand at all times. With such features as automatic tracking, worm gear drives, and computer interfaces that allow remote operation of the device from (usually warmer) indoor command posts, these are sometimes perplexing instruments to operate.

Lawnchair Nirvana: Meade's LX-200 Schmidt-Cassegrain telescope, fully loaded

As with binoculars, the ability of telescopes to efficiently gather light is paramount. This ability is directly related to the diameter of a telescope's objective mirror or lens, and its magnification power, which is determined by the distance between it and the eyepiece or focal point. Therefore, by increasing the focal length and interchanging a wide variety of wide-angle to close-up lenses, greater levels of magnification can be achieved.

In order to increase focal length while at the same time limiting the overall dimensions of the instrument (so you don't end up with a giant, ninety-foot-long tube), a Cassegrain telescope, named after the

Frenchman who devised it, uses a curved secondary mirror, which reflects light collected by the concave primary mirror straight back down the tube through a hole in the center of the primary mirror. This light-echoing effect of using the two mirrors causes a dramatic increase in the telescope's effective focal length over its actual length.

These instruments often are best suited to *everything*. It sounds funny, but it's true. This is the ultimate in Lawnchair devices. Using a Schmidt-Cassegrain telescope, together with a home computer and a host of photographic equipment, many Lawnchair Astronomers are doing things that only a decade or so ago were reserved for professionals.

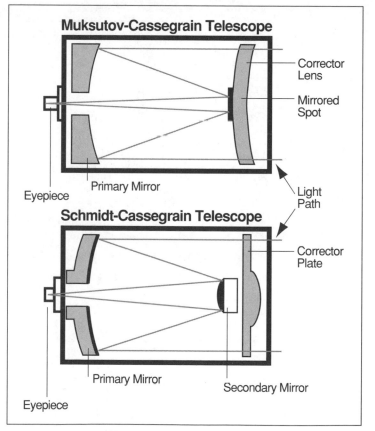

Schmidt-Cassegrain Telescope Design

ACCESSORIES
....

Lenses, coatings, filters, finder scopes, battery packs, stands, and mounts are just a few of the many options you'll be presented with during a telescope purchase. As mentioned earlier, an important element of ensuring a successful astronomical observation is the quality of the mount or stand you use in tandem with your telescope or binoculars: wooden, metal, old-fashioned, generic, or brand-name, each have pros and cons to consider.

Stability is what you are after in a stand or mount. An adequate mount will hold your telescope completely still so that even at its highest magnification, the image you see will remain clear. Most top-quality instruments are packaged with appropriate mounts, while lower-quality brand-name telescope makers sometime cut corners here. See the Buzzwords section for more information on the variety of mounts that are available today.

Buzzwords

In order to project that seasoned Lawnchair persona, use of the appropriate buzzwords is essential. Below is a sampling of what the well-informed Lawnchair Astronomer will soon be saying to his victims…er, friends and acquaintances.

★ **Aperture.** This determines how telescopes are classified. It is a measure of the diameter of a telescope's main objective lens or mirror. Seeking that ultimate thrill and to quell *aperture envy*, Lawnchair Astronomers have been known to spend incredible amounts of time, energy, and money to outsize the Jones's mighty tool!

★ **Chromatic Aberration.** Using lenses to bend light causes different wavelengths to bend at different angles. Often, in inexpensive instruments, stars will seem to have a bluish or purplish tint to their outer rings. This aberration is a result of the blue-violet portion of the spectrum being bent a little more than the rest of the light entering the lens. Reflecting telescopes (mirror based) do not suffer this fate. In refractors, a wide variety of glasses and crystals, often composed of fluorite, can be used to adjust for the aberration. You'll find these instruments advertised as "achromatic" telescopes, or "apochromatic" for those with nearly perfect optics.

★ **Collimation.** The optics of a telescope, generally reflecting telescopes, can be tuned to perfection allowing for the best possible, clearly focused image. If your telescope's optics are out of collimation, then an adjustment is usually required by moving either the primary or secondary mirrors slightly, until the optics are "true" — lined up and pointing accurately toward each other. Simple — though sometimes complex mechanisms — to make these adjustments accompany most telescopes.

Eyepiece Types

★ **Kellner.** These, like most eyepieces, come in a variety of sizes, are generally inexpensive, and are usually bundled with department store telescopes. They offer poor eye-relief (a measure of how comfortable they are to see through) and narrow fields of view. There are better-performing lenses available.

★ **Orthoscopic.** A good lens at a good price; the Erfle design, at the top of the line, can be quite expensive but provides fine wide-field views (a large piece of sky as opposed to a close-up look at a specific object) and excellent eye-relief; the Plossl, known as the best all-around eyepiece, comes in both moderate and expensive ranges, and with both wide- and narrow-field capabilities.

★ **Barlow.** A Barlow lens is a device that, by lengthening the effective focal length, has the effect of increasing the overall magnification. A 2X Barlow will double the magnification, a 3X will triple it. Modern Barlows are of a higher quality than were those of the not too recent past. When selecting eyepieces, remember that a Barlow can complement your collection and allow you to approach and often surpass your instrument's magnification capabilities.

Mounts

★ **Altazimuth Mount.** The traditional tripod mount. Allows movement in two directions: parallel to the ground (azimuth), and at right angles to the ground (altitude). Useful for terrestrial observations but significantly less useful for astronomical use, where an equatorial mount is preferred.

★ **Equatorial Mount.** Based on the celestial equator, this telescope mounting technique is designed to move your telescope across the sky aligned to specific, equatorially fixed points, in order to accurately locate and plot coordinates of celestial targets. For instance, observers using an equatorially mounted telescope in the northern hemisphere would adjust their mount according to their particular global latitude — a telescope at 35 degrees north would be mounted at a different angle than an observer at 45 degrees north — though both observers would align their instruments to the North Star for purposes of finding targets via an agreed-upon coordinate system. Employing measurements known as "right ascension and declination," an equatorial mount, combined with a motorized tracking mechanism, also allows tracking of celestial objects.

★ **Right Ascension and Declination.** All astronomical objects can be found by way of specific coordinates. Right ascension and declination are easily visualized by picturing the Earth within an outer sphere — a ball within a ball. By projecting the Earth's lines of latitude and longitude out onto the inside of the outer sphere, we can mark specific points in the sky and record unique coordinates for the immense variety of celestial objects. Declination is the celestial latitude and right ascension is the celestial longitude. In addition to their Messier or NGC designations, star catalogs typically include a star or deep-sky object's respective *Ra* and *Dec* coordinates.

★ **German Equatorial Mount.** These were the first equatorial mounts devised and are still the most common for small- to moderate-size reflectors and refractors. It relies on movable counterweights, which make them easy to balance but can also make them cumbersome to move. The telescope tube is usually held to this mount with encapsulating rings connected to a shaft upon which weights are fastened. This primary shaft rotates along the declination (up and down) axis. It is also cross-mounted along another secondary shaft, which is aligned to the polar axis. The polar axis is pointed at the celestial pole (the North Star), as is any other equatorial mount. In this design the use of weights and counterweights is necessary for smooth movement under high magnifications.

★ **Fork Mount.** This mount holds its telescope within two extended arms and allows the scope to swing up and down between them. A fork mount can be altazimuth or be made equatorial by simply adding an equatorial wedge. They are most commonly used with Schmidt-Cassegrain telescopes and are almost always set up equatorially.

The Universe on a Disk

When getting outside is just not possible, like when the weather isn't in a very cooperative mood, the Lawnchair Astronomer need not be thwarted. Via laptop or desk-model home computer, from the comfort of your very own living room, the universe can be yours to control, reverse, animate, speed up, and generally fool around in.

After a disappointing night out under an originally dark, perfect-for-stargazing sky that eventually became brightened and Moon-filled, you can now, with the aid of an astronomy program, easily (and quickly) keep track not only of when the Moon will be up, but of where planets, individual stars, and constellations will be at any point in time, and whether or not you can see them from your location (weather permitting, of course).

The ability to take a look at what's in the sky tonight, minus the clouds, from the same perspective as provided in your own backyard — or from anywhere else in the world — is just one of the many fascinating features you'll find in any of the dozens of offerings in astronomy and planetarium software available today. Through mail-order houses, on-line computer networks, or your local software retailer, you'll find an incredible variety of software, from easy-to-use sky projections to incredibly sophisticated astronomy databases. As with the purchase of binoculars or a telescope, it's best to shop around before buying, to find the appropriate software for your needs and hardware requirements/limitations.

Often advertised as "the best astronomy program ever," many of these programs also receive very positive reviews in the popular industry magazines. So how do you know which of them fulfill the claims made in their very appealing and slick ad campaigns? With a little investigation you'll be able to make an informed decision — and, in no time at all, you can be running your own indoor astronomical observatory!

If your computer has a modem, there are many fine "share-ware" programs available on both local and national bulletin boards and networks, and they're a great place to start. These programs will give you a taste of what an astronomy program can do. They often provide similar and sometimes better features than do programs that cost several hundreds of dollars more, so try them first. If they meet your needs, then you may be content with their sometimes limited flexibility and size. If not, you'll still have gained an understanding of what to look for in the higher-priced packages.

One great feature to look for is the ability to turn on and off various superimposed learning and object-location aids. Among such options are horizon, ecliptic, azimuth, and ascension markers, names of stars and deep-sky objects, and constellation lines — all of which can be immensely helpful in learning what's up there in the real sky, which of course doesn't have such an accommodating feature. Toggling between viewing modes — seeing a constellation with its lines in, then not, and then in relation to nearby constellations — helps to burn the image into your mind's eye. This will really help to orient you in searching your chosen area of sky once you get out under the stars.

Besides these aids, many programs provide additional database-type information about selected objects, such as planets, comets, and Messier objects. In some of the higher-priced packages, several star catalogs and lists, as well as pictures of the bright clusters, nebulae, and galaxies are available as well. Finally, the current state-of-the-art in computer-aided Lawnchair Astronomy is the combination of computer and telescope. There are programs available today that allow you to remotely guide your telescope toward a selected target, then photograph and electronically record it, just like the pros!

There seems to be no limitation to what the ideally outfitted Lawnchair Astronomer can achieve! Even without this high-end capability, however, the power to look at any part of the sky at will, even on the coldest, snowiest winter nights, is enough to warrant a closer look at this great tool for both long-time and novice Lawnchair Astronomers.

An example of what you can see on your computer: A computer reproduction of the sky for a night in October 1994; on the opposite page is the same "sky," with overlays connecting and naming the constellations.

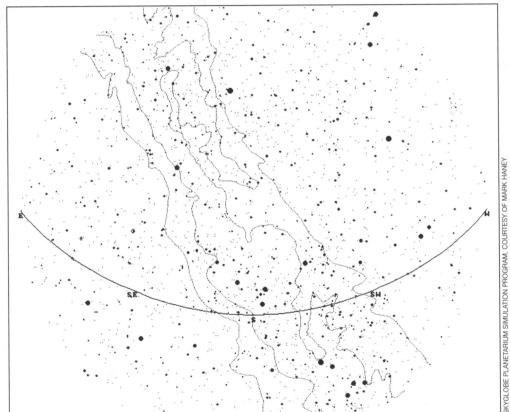

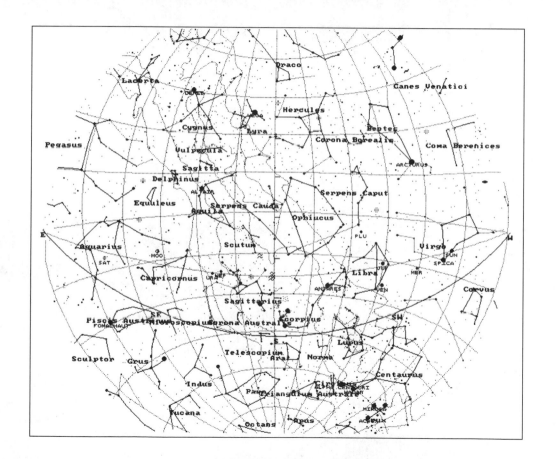

Astrophotography

Astrophotography is another fun activity that the well-equipped Lawnchair Astronomer can partake in. Because of the mechanics involved, though, this is a pursuit best suited to hard-core enthusiasts. But don't get me wrong: Anyone can create fascinating astrophotographs. Today, many of the pictures being published in the astronomy journals originate from backyard lawnchairs.

The basics include: a camera (35mm is fine), a special adapter to attach your camera to the telescope, which can be found in most telescope or camera shops, and, of course, a telescope. Actually, the latter is not entirely necessary. Simple "star-trail" photographs can be produced simply by pointing a camera toward the North Star and leaving its lens open for several hours. This is how the picture here was taken. These multicolored streaks of light are images of stars smeared across the film: a direct result of the Earth's rotation. This is the same reason that objects float across your field of view when you look through the telescope. It's not the object that is moving, it's you!

Using more expensive telescopes, those with the ability to track an object by way of a motorized mount, is absolutely necessary to produce the type of pinpoint images seen in photos found throughout this book. The reason is that since

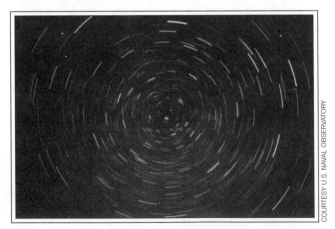

COURTESY U.S. NAVAL OBSERVATORY

A star-trail photograph by the folks at the U.S. Naval Observatory

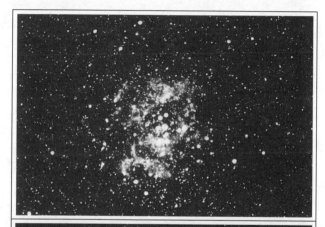

many of these objects are quite dim, to take their picture requires exposing the film for extended periods of time. Once the image is centered, in order to capture it clearly, it must be tracked perfectly to end up with such fascinating photos.

Simple pictures of constellations and planetary conjunctions (when two planets or a planet and the Moon travel close together) can be taken with just a camera mounted on a tripod. The requirement here is a "fast" film that can capture the points of light quickly without the images being smeared as in star-trail pictures. Advanced astrophotographers delve into the world of exotic "cold film" formulas and developing techniques. There are also several books that deal with this fascinating subject. A trip to the library will provide a great starting point for anyone interested in putting a star on celluloid!

Lawnchair astrophotography can yield such arresting images as these of the Rosette Nebula (top) and the Dumbbell Nebula (bottom).

SEE FOR YOURSELF: TOURING THE NEIGHBORHOOD

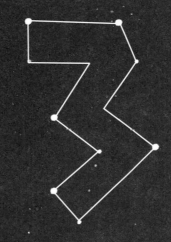

The Ecliptic—Keys to the Solar System

The ecliptic is where all the action is. As the very first Lawnchair Astronomers learned, there is a definite pattern that celestial objects follow. The Sun, the bright planets, our Moon, and twelve of the eighty-eight constellations follow a common path across the sky from east to west. This path migrates south in winter and then north again in summer, an effect of the Earth's seasonal tilting.

You'll come to recognize this path across your own backyard skies as you spend more and more time out on the old Lawnchair. Depending on the time of year, this rising and falling arc will either be low to the southern horizon, as in winter, or almost overhead, as it is during the summer months. Here is the path along which you will come to easily find the planets. All the exterior and interior planets travel along this path during the course of a year, unless they are traveling with the Sun. (As the planets travel along their paths around the Sun, they occasionally travel behind the Sun, from our perspective, and therefore are unavailable for us to see. This is where the term *traveling with the Sun* comes from.)

As seen from here on Earth, planets circling the Sun are absent from our skies during their trip behind or near the Sun. At rare though predictable intervals, from our vantage point, the Sun and Moon cross paths along this plane, providing us with spectacular eclipses of the Sun. Occasionally the Earth gets into the picture and it casts its shadow across the face of the Moon, which is called a lunar eclipse.

Generally, on any clear night, a Lawnchair Astronomer will at some point have the opportunity to see one or two of the major planets. Known as the naked-eye planets, even our distant ancestors knew that these five wandering rogues were not like other stars. Mercury, Venus, Mars, Jupiter, and Saturn can be seen without the aid of binoculars or telescopes if they are "up." The remaining three planets, Uranus, Neptune, and Pluto, will require a little more experience before you can easily locate them. After all, they were each discovered fairly recently, astronomically speaking, through painstaking study of the sky and planetary mechanics. As a matter of fact, Neptune and Pluto were predicted mathematically long before they were ever found observationally. Astronomers found them by noticing the slight gravitational tug they produced in the orbits of some of the other planets.

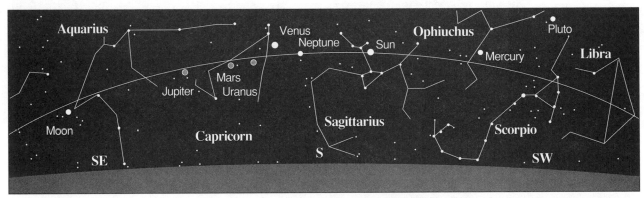

The planets, as well as the constellations of the zodiac, follow the ecliptic across the sky. The ecliptic rises and falls due to the changes in the Earth's axial tilt, which corresponds to the changing seasons.

The Sun

Many Lawnchair Astronomers are nocturnal. Enthusiasts often become extraordinarily pale, almost ghostlike in appearance, obviously a direct result of "the curse of the Lawnchair." It's not that they fear the Sun, it's just a little too bright for their taste.

Tragically, the only time these astronomers pay any attention to the Sun is when there is a solar eclipse. But, the truly liberated Lawnchair Astronomer knows that the Sun can be an awesome target to study at any time. Because it is so bright, though, stringent precautions are *absolutely necessary* before one can even attempt to gaze upon our fiery Sol.

Old King Sol: What can be said of this raging inferno that wouldn't make for an interesting read? To start with, the facts about this middle-aged, mediocre, medium-size star are well known by all in the astronomical community as well as by most of the general public. What might not be commonly known, though, are the various phenomena and processes associated with this engine that acts as our central heating system, sustaining all the life and myriad functions of our little, insignificant blue rock.

From a distance of a few hundred light-years away, a short hop in astronomical terms, Sol would look to an observer to be nothing special when compared with its neighboring stars. Although Sol might be a little brighter than many of the smaller dim stars in its neighborhood, it is just one of several local stars of similar stature, outshined only by a relative few giants on the outskirts of the group.

Here, back in the solar system, its effects are quite familiar to us. As recent as a few decades ago, though, we were not aware of its invisible force, the solar wind, which buffets all the planets, Earth included, with a broad range of radiation. Now, after several decades of observation from orbiting satellites (beginning with the first sputnik), a wealth of information about our home star has flowed into our data banks.

Taking approximately nine minutes to reach us, the Sun's energy, or radiation, in the form of heat and light, drives the myriad and complex systems here on Earth. Interplanetary space is also influenced by this outflow of energy in the form of excited particles, mostly electrons and ions. There are a few dramatic examples of how the solar wind can influence bodies in space. The most commonly viewed effects are the auroras seen around the North and South poles here on Earth. Another example is the way comet tails are blown back as they make their way in toward the Sun.

All manner of radiation originating from the Sun bombards the Earth's atmosphere constantly. These cosmic rays, such as those in the gamma and ultraviolet forms, strike the perimeter of the planet at incredible speeds. The majority of them, though, are deflected back into space. Hence the need to rise above the atmosphere to detect many of these exotic forms of radiation from both the Sun and other interstellar sources. On the other hand, as we have been learning of late, the ultraviolet portion of this radiation is increasingly finding its way to the planet proper because of depleted ozone levels in the upper atmosphere.

With the aid of a proper solar filter (see below and page 17) studying the Sun can be a very rewarding Lawnchair activity. Over weeks and months, the solar surface will treat you to sunspots, which appear in a wide variety of shapes, sizes, and locations, and then disappear. You can also use these spots to note the Sun's rotation. Watching the Sun and these dark, odd-shaped blemishes from a properly decked-out locale, say Margueriteville or the beach, can make for quite a Lawnchair afternoon. Don't forget the tanning lotion!

Magnified to the extreme, a view of the Sun's outside edges, or corona, can provide awe-inspiring glimpses of the fury of solar flares, the bright, erupting plumes of solar material that explode thousands of miles up into Sol's atmosphere. Silhouetted against the dark background

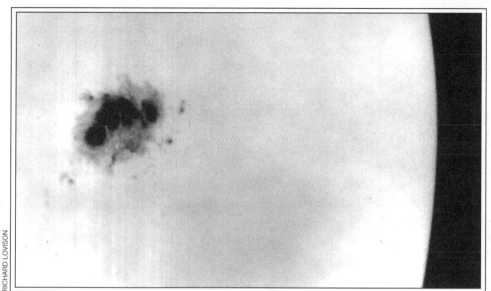

RICHARD LOVISON

A sunspot captured by a Lawnchair Astronomer's camera

Only two materials are recommended for solar observation: #14 Welder's glass and aluminized Mylar. The latter is the most preferred for telescopic filtering. Welder's glass is designed more as a shield than for fine optical performance. Aluminized Mylar, which you can purchase from most astronomical equipment dealers, comes in sheets, or already mounted in specific frame sizes to fit a wide variety of instruments. Aluminized Mylar deflects most of the Sun's energy, allowing safe levels of solar energy to enter your telescope, where you can magnify the image as you would any other astronomical target. Since aluminized Mylar comes in a variety

of space, this sight has thrilled Lawnchair Astronomers for centuries.

The most important thing to remember, though, is that looking at the Sun without adequate filtering will permanently damage eyesight.

49

of thicknesses, you might sometimes need to double, or even triple, it up to increase its effectiveness and, therefore, your safety.

An important note: The proper way to use a solar filter is to cover the main objective lens of the instrument, not the eyepiece. Attach it to the front part of your telescope or binoculars, by taping it either with duct tape or another nonporous, opaque material. In the case of commercial filters, they are usually fitted to snugly encapsulate the front end of your instrument. The idea is that no light must be allowed to enter the telescope tube. It must be completely covered. The Sun's unfiltered light, amplified by the optics of your telescope, could melt, warp, and destroy those very same optics.

Never use eyepiece-covering solar filters. They do not protect the optics of your telescope. For almost the same reason you wouldn't expose your eyes to the Sun's light, the lenses and/or mirrors in your telescope are just as negatively affected. With the kind of energy coming from the Sun, it doesn't take long for this telescopically amplified energy to damage your instrument. Often, lens-covering filters get so hot that they can crack or even shatter. And it's too late at that point. Use only filters that can completely cover the front part of your telescope or binoculars.

Lastly, a great benefit of daytime is that it can be done anywhere the Sun shines. Often, local astrogroups set up their toys...er, telescopes at sidewalk sights, usually near a mall with abundant foot traffic. If you spot one of these curious sights, be sure to introduce yourself and take a peek at ol' Sol. You just might find that you have a lot in common with these sidewalk astronomers — They've got lawnchairs at home too!

The Moon:
Lighting a Lawnchair Near You!

Seeing the Moon through a quality instrument can be breathtaking. The craters, hundreds of them, are singularly striking features in themselves. The mountaintops, where sunlight peeks through and spreads shadows over hundred-mile basins, are quite apparent in observations during either the waning or waxing (first and last quarter) periods. A full Moon, though, is not the best time to study our nearest neighbor. Seeing the terminator, the point where night and day occur on the lunar surface, can only be accomplished when the moon is *not* full. This is where shadows can be used to determine the height and width of the aforementioned lunar features. The best time for lunar observing is during the first and last quarter, and half moon phases.

The Moon is new when it travels with the Sun. In other words, when the Moon is up and in the same vicinity of the sky as the Sun, it naturally rises and sets with the Sun. On these evenings, minus the Moon's obstructing glare, deep-sky Lawnchair Astronomy is at its finest. A few days past new, and the familiar crescent is visible in the west just after sunset. On successive evenings it climbs higher into the sky and grows in size. Once full, it is rising in the east as the Sun sets. In reverse, over the next half of the month, it shrinks and rises later and later at night, till in early morning it is once again rising and, later on, setting with the Sun.

The phases of the Moon take approximately twenty-eight days to complete. Blamed for everything from causing lunacy when full to governing the menstrual cycles of females of every species and the Earth's tides, the Moon has long fascinated the Lawnchair Astronomer.

With binoculars or a small telescope (big scopes are not best suited for the Moon and planets) the Moon presents a wealth of views. In a wide angle, while waning or waxing (say that fast three times) you'll see details of the Moon's crusty surface along the terminator. In close-up views, the largest craters are simply magnificent to behold. Along the terminator, a close-up will let you peek across the day-night separation and see streaks of setting-sun light cross over into the darkened side of the lunar landscape. Here, also, you can get a feeling for the height of the lunar mountain ranges and crater walls, by comparing the length of the shadows they cast out onto the bright side of the terminator. It's a really nice place to visit, and best of all, you can do it from your own Lawnchair.

RICHARD LEVISON

The Moon in its various phases (from l. to r.): A wide crescent, quarter phase, and full. A full Moon, though dramatic and romantic, doesn't make as good a viewing target as a side-lit Moon in a lesser phase.

Mercury

Not the smallest of the planets, it is larger than our Moon and the planet Pluto. Like the Moon, it is scarred by prehistoric interplanetary impacts. Until the mid-seventies, very little was known about the innermost planet. The Greeks were confused by its odd appearance (it has phases such as those we see on the Moon and Venus) and by the retrograde motion it displays.

As a planet circles the Sun, from our perspective over several nights' observations, it can seem to stand still and not progress along its path on the ecliptic, and even go backward in the sky. Watching a planet at the same time each night over the course of several nights or weeks, you will normally notice that it has moved a little farther along the ecliptic. And that is because it is circling the Sun and actually moving through the solar system. But, occasionally, from our perspective, as a planet turns the corner, so to speak, it seems to stop and travel backward over time — detectable over a few nights with instruments or,

with the naked eye, over the course of a couple of weeks.

Mercury is one of the brightest planets, a result of its proximity to the Sun. But because of that relationship it is also quite hard to view except for a few times during the year when conditions are just right. It is best seen from Earth during aphelion.

Aphelion is when a planet is at its farthest from the Sun and from our perspective rises or sets at its farthest point away from the Sun. Also, depending on which side of the Sun a planet is — traveling toward or away from it — a planet will either appear in our morning or evening sky, rising before the Sun does in the east, or setting after the Sun does in the west. In the case of the inner planets, Mercury and Venus, whose orbits fit within that of the Earth's, there is a limited amount of time they can be seen in the morning or evening skies.

And in the quest to Royal Lawnchairhood, one

MERCURY: PECULIAR QUICKSILVER

Closest to the Sun, Mercury is still over thirty-six million miles away from it. One of the most peculiar aspects of Mercury is its rotation on its axis and how it relates to its orbit around the Sun. During two trips around the Sun, Mercury only rotates on its axis three times. In essence, a day and a half on Mercury is equal to a year! Each rotation takes 58.6 Earth days and it completes one revolution around the Sun in 88 of our days.

Another peculiar feature that's been uncovered about Mercury is that it has two hot spots, where, during perihelion (closest approach to the Sun) Sol sits directly overhead in the Mercurial sky. The mean temperature on the surface at this point during the day reaches 660 degrees, while at night it cools to a frigid -270 degrees.

Finally, Mercury is without a partner. It has no moon. It does sport a rather large impact crater, though, called Caloris, which has been measured at over eight hundred miles across. Could it have been caused by a one-time partner? Also, like most airless bodies in the solar system, Mercury is covered by debris or regolith from ancient impacts. These, unfortunately, are not features available to Lawnchair-grade instruments.

of the premier trophy pieces is to have seen an optimally contrasted view of Mercury, silhouetted against a relatively dark sky. To do this, you must aim for those rare occasions when these perfect conditions exist. This information is available from local observatories and is also published in *Astronomy* and *Sky & Telescope* magazines.

To spot Mercury, it is absolutely necessary to be where you can get a good view of the appropriate horizon, preferably from atop a hill or mountain. The bright object that you may glimpse in the haze above where the ground meets the sky will be this elusive sibling. At its furthest separation from the Sun, Mercury can just barely be seen. To complicate matters, looking through the Earth's atmosphere toward the horizon during or just after sunset is far from ideal for studying any astronomical target.

In 1974 and 1975, the *Mariner 10* spacecraft made several fly-bys of Mercury, and astronomers finally got their first close-up look of the Sun's nearest planet. They discovered that Mercury's surface has two distinctive features, which even today generate debate as to their origins. One portion of the Mercurian landscape is saturated with impact craters, while the other is comprised of smooth, rolling plains. Theories abound as to the origin of these different surface features. Some propose that the plains are young volcanic lava fields, while others think that these areas represent the planet's original surface. The wealth of data collected by the *Mariner* mission is still being researched today, and there is still more to learn about this, our neighbor's neighbor.

So, if you like the thrill of the hunt, the fleet-footed "messenger of the gods" makes a great quest for either early-evening or predawn adventures. It's not everyone who can say they've seen this planet in person. But with some persistence and excellent weather conditions, you could add this trickiest of the major planets to your Lawnchair trophy wall.

Venus: Love Goddess with the Hots!

Venus is an ideal place to visit from your Lawnchair, because you surely would not want to live there. The planet that most closely resembles the Earth in size is a little closer to the Sun — but in astronomical distances it's just a stone's throw away from Earth. Venus, named after the Roman goddess of love, has long been a sight that captures the imagination.

Since the earliest of times, Venus has been a riddle. Even today, with the great strides in education and technology, the appearance of this brightest planet, shimmering on the horizon, sometimes causes great consternation. UFO reports abound when Venus pops out of the twilight and grows in brightness as the sky darkens. Hugging the horizon as it does, like the Moon, it creates an optical illusion in which it sometimes seems to travel along with you as you ride in a car.

Venus never gets very high into the sky, or so it seems, because of the same mechanics that are involved with viewing Mercury, the other "inner planet." But unlike Mercury, Venus spends a slightly longer time in dark skies, since it is that much farther away from the Sun. As the second planet out from the Sun, it is interior to our orbit and therefore rises and sets with the Sun, as Mercury does.

Venus also goes through phases. Just as with the Moon and Mercury, you, too, can see this phenomenon with a pair of binoculars or a small telescope. By watching Venus over a few weeks or months, you'll notice the phases changing or cycling from first quarter to half to last and then to new again. The reason we witness phases of the Moon and inner planets is because of our relationship to their orbits and to the Sun. At each different stage, we see the portions that are lit by the Sun.

From the Lawnchair, unfortunately, nothing more than the phases and shimmering tops of miles-deep cloud layers are visible. These clouds

NASA

Venus Unveiled: This computer-simulated view of the surface of Venus was based on the data collected by NASA's orbiting Magellan *radar.*

are responsible for Venus's bright appearance, reflecting an enormous amount of the Sun's light. What lies beneath, though, is worthy of inclusion

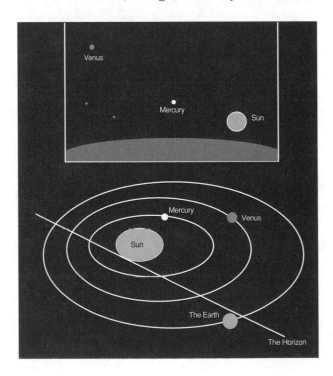

here, since the planet has mystified both professional and Lawnchair astronomers for millennia. Venus's obscured surface consists of wildly volcanic terrain. Dry, black, and charred, the surface of Venus is unlike anything concocted in the vivid imagination of science fiction writers past and present. The recent mission to map with radar this planet with a hellish surface provided as many new questions as it answered. Venus seems to have been recently resurfaced within the last several hundred million years; compared to surfaces of the Moon and Mars, which have been calculated at over several billion years old, Venus is considered "new."

The unmanned probe *Magellan* has completed its multiple radar mapping passes around Venus. The results and photographs of the mission are now available for research, and so far have resulted in some startling finds. Scientists were stunned to find that Venus sports relatively

The celestial mechanics behind the horizon-hugging ways of Mercury and Venus: Because these two planets are interior to the Earth's orbit of the Sun, they will always set soon after or before the Sun does.

few impact craters. And the few that were found, one of which is over two hundred and fifty miles wide, are in themselves peculiar compared to what is commonly witnessed on other planetary bodies. Aside from the craters, recently crusted lava flows, like those seen here on Earth, are the other main feature of Venus's surface. So far, though, no active volcanoes have been detected, but additional observations are planned and targeted on the areas of greatest promise.

Looking at Venus is often made even more interesting when it shares the spotlight during occultations (planetary and stellar eclipses) and conjunctions with the other celestial bodies. Brighter than any other object in the sky, save for the Sun, Venus always manages to make an impression on the interested observer. Spread out the Lawnchair and play voyeur to the goddess of love tonight!

GREAT MOMENTS IN LAWNCHAIR ASTRONOMY

The First Heliocentrist

In ancient Greece, the geocentric (Earth at the center) view of the universe postulated by Aristotle (384–322 B.C.) was accepted by almost all, except for a careful astronomer by the name of Aristarchus of Samos (fl. ca. 270 B.C.), who not only proposed a heliocentric (Sun at the center) view, but also proclaimed that the stars were at much greater distances than was believed at the time. Sadly, his views were not accepted by his contemporaries. Though no substantial proof exists, it is rumored that jealousy over his extensive collection of fine Grecian Lawnchairs was at the root of this alienation.

Mars: Where Are Them Pesky Martians?

Mars, whose warm reddish tint is easily apparent to even the unaided eye, contrasts sharply with the bright, highly reflective white cloud-tops of Venus. Mars has no obscuring layers of clouds, though it has been known to entertain energetic sand storms. The red planet, perhaps the most fictionalized place in the universe, is quite an easy and exciting target from the Lawnchair. Depending on where it is in its cycle (how close to the Earth or Sun it is), a pair of binoculars or small telescope will reveal many interesting features. Occasionally, when passing relatively close to the Earth in its orbit every few years or so, Mars presents its best (largest) face for backyard observation.

Throughout history Mars has been an enigma. Earliest Lawnchair Astronomers were fascinated by its color and its seemingly undecided (retrograde) path across the sky. Its red-eyed nature earned it the name Mars, for the god of war. Seasonal growth and shrinkage of the polar ice caps, as well as a wave of darkening that appears to sweep the Martian surface during each hemisphere's spring season, regularly star in the eyepieces of telescopes belonging to the most astute Lawnchair Astronomers.

Percival Lowell (1855–1916) was just such an individual. Perhaps encouraged by the fantasies of the era's fiction writers, he saw canals that he was sure at one time carried water to the Martian cities after being produced at the Martian polar ice caps in giant melting stations. Lowell established the Lowell Observatory in Flagstaff, Arizona, where he used both 18-inch and 24-inch refracting telescopes to study the Martian homeland. After taking the first good photograph of the red world in 1907, he proclaimed that his pictures proved that the canals were indeed Martian-made.

Just like the Earth, Mars is tilted on its rotational axis and enjoys seasonal variations as a result. It's theorized that temperatures may have been warmer on Mars in ancient times. Perhaps

NASA

The Martian surface, as captured by NASA's Viking II *lander (note: no welcoming Martian party)*

water that today only exists frozen in the polar ice caps, flowed at least seasonally in and out of the canals (dried-up river beds) that Percival Lowell saw. What he never saw were Mars's two orbiting moons, Phobos and Deimos, each unfortunately too small for detection by Lawnchair instruments.

Mars's oceanless surface area is almost equal in size to all the land areas on Earth, though the planet as a whole is only half the total size of Earth. Its day is almost identical to ours, only thirty-seven minutes longer, though a Martian year (one revolution around the Sun) lasts twice as long as an Earth year.

MARS: COLD DAYS, COLDER NIGHTS, AND A HUGE VOLCANO

Several robotic missions have pretty much solved any of the remaining mysteries about our sister world. Additional visits have concentrated more on geologic and meteorologic features. Today, still interested in looking for life there, albeit in a microbial and fossilized form, several plans, both manned and robotic, are again afoot.

The now apparently defunct *Mars Observer* was scheduled to begin relaying information from Mars in August 1993. If it had worked, it would have been the twelfth successful mission to the red planet. Unfortunately it did not live up to the successes of its predecessors. In the news the words you saw most often were *it broke*. That's about all anyone could say. The

From a locale where a lawnchair can be easily and comfortably utilized, whether for morning or evening visits, seeing Mars and its sometimes dramatic features make for great Lawnchair accomplishments. Dynamically different over time, views of the red planet are constantly revealing something new. Look for a lightening of the disk's edges through a pair of binoculars, and in telescopes, small and large, try to notice the darkening or changes over time of the different hemispheres. As the resident Lawnchair expert, your family and friends will listen, spellbound, to your tales of conquering the Martian home world.

more colloquial term among Lawnchair Astronomers is similar to the phrase *excrement occurs*.

Mars has an atmosphere composed of carbon dioxide, though it is quite thin. And the temperature on a typical Martian day can start at a cool 117 degrees below zero Fahrenheit (-83 Celsius), and by noon reach a balmy 27 degrees below zero. Makes you realize just how comfortable that backyard Lawnchair really is after all!

Mars also endured a volcanic period about a billion years ago. As a matter of fact, the largest volcano in the solar system, called Olympus Mons, extends a whopping twenty-seven kilometers (seventeen miles) above the Martian landscape. At three times the size of our Mount Everest, this behemoth mountain sits on what is known as the Tharsis Plateau, a large terrain feature dotted with several other volcanically produced peaks. The largest Lawnchair instruments, the cannon-sized 12-inch to 17.5-inch instruments, are perhaps the only way to glimpse these features. And even then, only under ideal conditions.

Jupiter: Giant Beachball in the Sky

This giant gas-and-cloud planet offers Lawnchair Astronomers everywhere plenty to see. Its revolving cloud tops are quite interesting, as are its four large, orbiting moons. With a small telescope or binoculars held steady, you can see these four large moons circling Jupiter, each in a different locale every night. Jupiter's famous cloud bands also contain the Great Red Spot, a turbulent atmospheric storm that is larger than the Earth. Looking at Jupiter through small and medium-size telescopes, over an evening's outing you will see changes as they occur in the cloud bands. By going back to Jupiter every half hour or so during an evening's foray into the night sky, you can get a feel for how dynamic this giant world is.

Jupiter's diameter is greater at its equator than it is from pole to pole. From the perspective of the Lawnchair eyepiece, the planet will seem to be slightly flattened, a not quite round sphere. Even though it is ten times bigger than Earth, it completes one entire rotation, a Jovian day, in less than ten Earth hours. This rapid rotation of Jupiter on its axis creates a centrifugal force that is responsible for the flattening at the poles.

Except for when it travels behind, or with, the Sun, Jupiter is usually available in either evening and early morning skies for sometimes more than a year at a time. Jupiter is best viewed when its and Earth's orbits place the planets near each other. At that time, its few features are apparent even with the smallest Lawnchair instruments.

There are sixteen satellites circling Jupiter, almost a miniature solar system. As a matter of fact, during the primordial, formative period of the solar system, Jupiter fell just short of the amount of mass needed to compress itself and eventually ignite into a star. Many stars seen within the Milky Way are indeed members of binary and often multiple star systems rotating around each other or a common center of gravi-

NASA

Jupiter

Jupiter's four largest moons are known as the Galilean satellites, after their discoverer, Galileo Galilei, who first saw these four large moons in 1610. Io, Europa, Ganymede, and Callisto each have their own unique properties.

Io is continually squeezed in and out like an accordion by the strong gravitational forces of its parent. This tidal flexing effect generates extreme internal friction and heat in Io's core. As a result Io has volcanoes, noted by the *Voyager* missions.

Europa is predominantly a rock body with a smooth outer casing made of ice. Also affected by Jupiter's intense gravity, Europa is racked by tidal flexing, and its exterior is constantly being resurfaced. Internal heat and volcanic processes serve to melt its surface ice, forming a sort of ice-and-lava flow over broad regions of the large moon.

ty. Fortunately for us, this did not occur in our neighborhood. If it had, we wouldn't be here to discuss it!

Ganymede and Callisto are both cold and barren worlds. Rock cores make up almost half of each of their mass, with their other halves com-

NASA

The large Jovian moon Io is dwarfed by Jupiter's sphere in this Voyager II *image.*

prised of outer ice shells, not unlike many frozen bodies found throughout the solar system.

In any case, all that is available to Lawnchair Astronomers are the changing positions of the largest moons, and perhaps an occasional glimpse of one or two of the larger minor moons. And this only under the most ideal conditions.

JUPITER: STRONG MAGNETISM, THIN RINGS

Jupiter produces the strongest magnetic field of all the planets. Extending out farther than ten times the planet's radius, this magnetic field is evident in strong bursts of radio noise detected here by radio telescopes. The source of this field may originate from Jupiter's exotic interior. Hydrogen and helium in their normal gas form make up most of the planet Jupiter. Its rocky core, which is at least ten times larger than the Earth, is surrounded by a thin, though highly compressed, metallic-liquid ocean. This is a direct result of the extreme pressures on this huge planet, which is almost as large as the Sun.

Once astronomers visited Jupiter, although remotely via robotic spacecraft (the *Voyager* missions), they discovered that Saturn isn't the only planet that has rings. Jupiter possesses a narrow system of rings composed of tiny rocks and dust particles very similar to, though not as extensive as that of, the other Jovian-size planet, Saturn.

THE LAWNCHAIR EVENT OF A LIFETIME

In July 1994, several pieces of a captured comet crashed into the gas giant Jupiter. This never-before-witnessed event provided both professional and Lawnchair astronomers a great opportunity to see what happens when celestial bodies collide, an event that has occurred thousands of times in the five-billion-year history of the solar system. There is substantial evidence (impact craters) of both ancient and recent (relatively speaking) collisions throughout the solar neighborhood.

Archeologists and astronomers have noted numerous prehistoric impact sites throughout the solar system. They are very apparent on our nearest neighbors, the Moon, Mars, Venus, and Mercury, and even here on Earth, where most impact scars have "grown over."

When these cometary fragments struck and exploded violently on the outward side of Jupiter, astronomers all over the world witnessed what was perhaps one of the most important astronomical events of the twentieth century. As a first line of detection, astronomers may someday use this same experience to track space debris bound for Earth.

Saturn: The Rings That Bind Us

Picture yourself in this real-life predicament. You've spent months planning a party for all the people you need to impress the most. In the middle of it all, between the pâté and the bobbing for kiwi, disaster strikes as the power goes out.

For anyone else the night may seem ruined, but not to the quick-thinking Lawnchair Astronomer. You look over to that comforting site in the middle of the backyard, where you know that in minutes you'll once again be the host with the most. Just mention the word *Saturn* and immediately guests stop in midsentence and the line begins to form at the eyepiece, as word spreads through the crowd that "you really *can* see the rings!"

The obvious star of any planetary show is Saturn. If seen only with binoculars, the magnificent ring system is still wonderful to witness, and even more so in larger instruments. With a small telescope the ringed planet is quite impressive, especially once you begin to increase mag-

nification levels to their limits. With large backyard telescopes, the major separations in the ring system are absolutely possible to discern, though only under the most calm and dark skies.

Saturn could take up a whole night's worth of Lawnchairing. Starting with a wide-angle eyepiece, try to see at least two of her large moons — they sparkle outside of what might seem the natural orbital position (where you'd think that they'd reside) along the same plane as the rings. On a moonless night, depending on your instrument and the sky conditions, you may possibly glimpse four of Saturn's largest satellites: Titan, Tethys, Rhea, and Dione. Saturn sports an incredible twenty orbiting moons in all. The largest is Titan, bigger than Mercury but not quite the size of Mars.

Once satiated with that view, try a higher magnification eyepiece. If the seeing is really good, look for both the major and minor divisions in the ring system. This is the brass ring of plan-

Saturn

NASA

etary observing. Cassini's division, named after its discoverer, another Lawnchair Astronomer with really great eyes, is the outermost division in the ring system. It is also the largest gap visible to earthbound Lawnchairs.

Another division in the rings visible from your backyard is called Encke's division and will require, at a minimum, a steady (rock solid), mounted 8-inch telescope and better than perfect seeing conditions. Only attempt this during clear, new Moon periods, when the air is still. Look for, at best, a thin, gray to black line, substantially smaller than the large Cassini's division. If you score either of these, you'll have earned the Lawnchair Eagle-eye Merit Badge. Wear it proudly on your most treasured lawnchair.

Next, to enhance the subtle planetary markings, clouds, and belts similar to Jupiter's, try using a blue or green eyepiece filter, which under good conditions should show some remarkable Saturnine features. A yellow or amber filter will also highlight some subtle features of the ring system as well. Try as many combinations as you have time for. Each will bring out a different aspect of this world.

In the late seventies and eighties, cameras aboard *Voyager*s *I* and *II* sent home pictures of our solar system neighbors that had never before been seen with human eyes. During extended fly-bys and gravity-assisted encounters with the gas giants — Jupiter and Saturn — the *Voyager*s helped to resolve the long-standing mysteries about what Saturn's rings were made of. Astronomers learned that the ring system is composed of millions of boulders, rocks, pebbles, and ice-covered dust particles circling the planet in intertwined or braided strands. The photographs revealed that there are thousands of rings, from about 4,300 miles (7,000 kilometers) to 46,000 miles (74,000 kilometers) above the Saturnine cloud tops.

As the night wears on and some of your guests begin to notice the Sun rising in the east, you'll have learned that the Lawnchair bug is easy to spread. And, after all, your duty as a Lawnchair Astronomer is to do just that. You'll be rewarded for your efforts by your new stature as the ultimate party animal. In your best *Blazing Saddles* voice, repeat after me: "Electricity, we don't need no stinking electricity!"

Uranus, Neptune, and Pluto... or Is that Pluto and Neptune?

To really "see" Neptune and Uranus will require at least a pair of binoculars, or better yet a small telescope. In a preferably moonless sky, Uranus can be seen with the naked eye if viewing (weather) conditions are exceptional. A telescope will reveal its bluish green tint, as well as a dark-ening around the edges of its disk. Because of its distance to the Earth it appears rather small, though in reality it is a substantial world some four times the size of our big blue planet. Seventh out from the Sun, it was first discovered by Sir William Herschel in 1781.

URANUS COULD HAVE BEEN A GEORGE

When Uranus was first found, its discoverer, Sir William Herschel, wanted the planet to be named Georgium Sidus or Georgian Star, after the King of England at the time, George III. Many French astronomers referred to it simply as Herschel and it wasn't until later on when Johann Bode (1747–1826) proposed it be named Uranus after the mythological father of Saturn.

Neptune

Neptune, the eighth child of Sol, is the farthest of the gas giants. Very similar to Jupiter and Saturn, Neptune also possesses Earth-like characteristics. Its weather patterns seem to be quite active and its tilt resembles that of terra firma. Neptune, though, cannot be seen without the aid of telescopics at any time. Shining dimly at about magnitude 7.8, you will need at least a 6-inch telescope to capture this planetary target. In that field of view, you will see a fairly unspectacular sight. Just seeing this small blue disk some three plus billion miles away from your Lawnchair is quite an accomplishment.

Neptune

We've learned much about Neptune and its giant, gaseous neighbor, Uranus, though not by way of earthbound telescopic observation. In the latter part of the eighties the *Voyager II* mission knocked on the cloud-top front doors of these most distant planets with startling results. It had been suggested that perhaps Uranus had its own partial rings similar to those of Saturn's; *Voyager* confirmed those predictions and revealed that Neptune also had rings.

In addition to finding the rings of each planet, the spacecraft returned pictures of additional moons that had never before been seen. Uranus, once thought to possess five large moons, was discovered to be harboring ten smaller moons in very close orbits. Neptune, whose two moons Triton and Nereid have been known for quite a while, was also guilty of hiding a number of sibling satellites, six in all.

GREAT MOMENTS IN
LAWNCHAIR ASTRONOMY

Huygens and Picard

The refinements that Christian Huygens (1629–1695) made to the telescope of his day helped pave the way for an intense period of astronomical discovery. Huygens, who ground his own lenses and painstakingly studied the heavens, was also responsible for introducing a more accurate method of time-keeping, with his development of the pendulum clock. Additionally, he, along with Jean Picard (1620–1682), no doubt a distant ancestor of the starship captain, devised a method for making much more accurate angular measurements of the stars.

Pluto

Pluto, the most distant planet, is also the smallest. Predicted mathematically through study of the perturbations in the orbits of Uranus and Neptune, this tiny planetoid was first detected by its movement on telescopic plates or astrophotographs taken in 1930. Its moon, Charon, wasn't discovered till almost fifty years later in 1978. Similar to Titan and other frozen Galilean satellites, Pluto's surface is covered by a layer of methane frost and water ice. Its combined mass, even including that of its moon Charon, is still 450 times smaller than the Earth's.

Pluto's orbit is much more elliptic than those of the other planets and is tilted 17 degrees out of the ecliptic plane. At perihelion (closest approach to the Sun), Pluto is nearer the Sun than Neptune is. This is the reason you hear that at some times Neptune is the farthest planet.

Looking like a dim star, this is a target for only the most seasoned and well-outfitted Lawnchair Astronomers. At a very dim 15.1 magnitude (and that's at its brightest), Pluto is still absolutely possible to find with a sufficiently suited backyard telescope. Armed with up-to-date coordinates, something you can often find in *Astronomy* or *Sky & Telescope*'s monthly sky guides, thousands of Lawnchair Astronomers have accomplished just such a feat.

For some, seeing Pluto and all the planets in person is a sort of rite of passage. For Lawnchair Astronomers, it also means that it's time to move out into the galaxy and the universe at large, where jewels flourish in the thousands. Spread out the blankets and unfold the Lawnchairs: the adventure is about to begin.

STEPPING OUT INTO THE UNIVERSE

The Constellations—
Road Maps of the Skies

I am often asked, "How do you find those little fuzzies so easily?" Or, "How come I can never find anything that I want to look at?" Seems this is one of the biggest hurdles that keeps many Lawnchair pledges from becoming full-fledged Lawnchair Astronomers. But it doesn't have to be this way. There is a trick to it and it's really simple.

It's the same idea as when you have to find a certain place, perhaps in a town you've only visited once before, some ten years earlier. You remember almost nothing, except that you can vaguely recall eating in a restaurant on the way there with a very distinctive decor. Now years later, while driving around trying to find your destination in this foreign town, you come upon a traffic circle. It has five exits to choose from. You spend a few revolutions examining each, and suddenly, off in the distance on exit three, to your sheer delight is what appears to be...yes, it's a giant chicken-shaped eatery!

Similarly, looking out across the sky can sometimes seem overwhelming. But, in a nutshell, by recognizing the familiar, it can be quite easy to find the not so familiar.

Many of the line pictures we know as constellations — there are eighty-eight in all — originated with ancient Lawnchair Astronomers, who saw and recorded their folklore and myths in the sky. For today's Lawnchair Astronomer, these same sky murals are the equivalent of the giant chicken you'd use to find your way to that unfamiliar neighborhood.

Names like Hercules, Orion, and Andromeda come directly from the ancients' campfire stories. Before writing was invented, the way people would remember the stories that came down

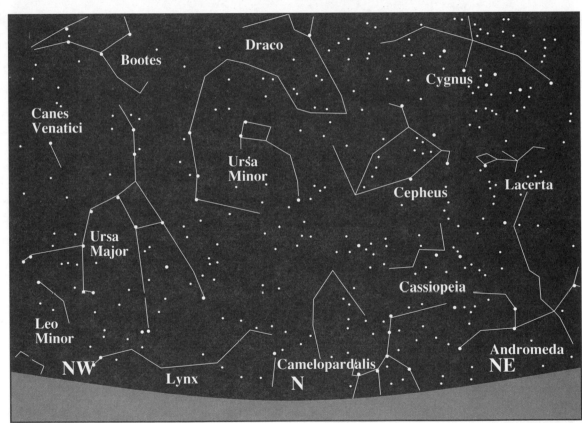

The original constellations, as envisioned by the ancients...

Bootes

Draco

Cygnus

Canes Venatici

Ursa Minor

Lacerta

Cepheus

Ursa Major

Cassiopeia

Leo Minor

Andromeda
NE

NW

Camelopardalis
N

Lynx

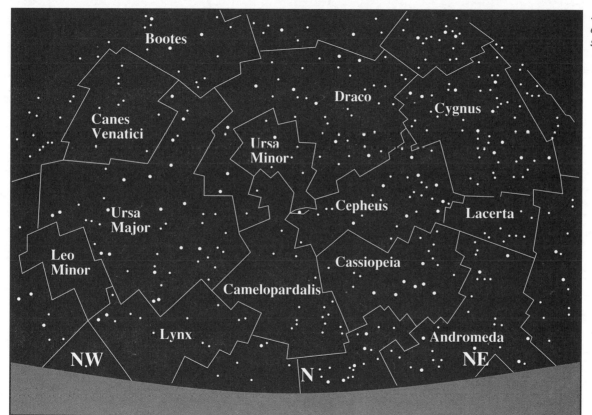

...and the modern adaptations of those same constellations.

through the generations was to tell them over and over again. At night they would use the starry background to visualize these stories and record them for a later telling.

The word *constellation* comes from the Latin meaning "together" and "stars." Many of the names are very old. Some of them come from over 7,000 years ago, when Mesopotamian farmers and shepherds imaginatively formed the connect-the-dot pictures in their nighttime skies that resembled animals they knew (the bull, the ram, the lion). The Greeks adopted many of these and added their own mythological figures to the growing list of star murals. Romans accepted the list but translated the names into Latin. By A.D. 150 Ptolemy listed forty-eight of the constellations he knew of in his text, the Almagest. As time went on, other astronomers added new constellations to Ptolemy's list. Some are even named for modern scientific instruments, such as the sextant, the compass, and the telescope. It does seem, though, that they've left out a very important one: the lawnchair.

Today, every star and deep-sky object can be found within the boundaries of one constellation or another. The modern adaptations of the constellations incorporate the ancient pictures, but mark them out in nontraditional straight lines and boxed-in areas laying out definite borders between the eighty-eight sections of sky. You're free to describe them in any manner you wish.

Lawnchair Astronomers use the constellations to identify the location of their favorite bright stars, or determine where the planets along the ecliptic will be, or even where they can find the treasured deep-sky treats that are so sought after. The best advice I can pass on to you is to become as familiar as you can with the constellations in your night skies. Begin with the ones you know well — the giant chickens — and work from there.

Many constellations consist of stars that are just at, or even below, the naked-eye visual limit. Use a star map or a computer/astroplanetarium program to study them. Look for the familiar bright stars and trace out their companions. It's really quite easy. And the darker the sky is, the easier it gets.

There are some constellations that can only be seen from the northern hemisphere and, conversely, some that can only be seen from the southern hemisphere. Year-round, in the northern hemisphere the constellations that appear to rotate around Polaris, the polestar or North Star, are known as the circumpolar constellations. From below the equator, though, those constellations are never visible. The reverse is true for southern and northern hemispheres with regard to constellations associated with the southern polestar, Sigma Octanus.

The twelve constellations of the zodiac travel along with the Sun, Moon and planets on the ecliptic, and can be seen from both hemispheres. From there, the remaining constellations are divided into two groups: those north of the zodiac and those south of it — the northern and southern constellations.

Each month, as the Earth revolves around the Sun, the night sky reveals new sets of constellations. Over the length of an evening, a diligent Lawnchair Astronomer can sneak a look at the upcoming season of constellations as they rise later in the evening and early morning, up until the Sun rises. Constellations visible at four A.M. today, won't be visible in the evening skies for several more months. Conversely, constellations setting in early evening, won't reappear in morning skies for some time as well. So in addition to northern and southern groupings, constellations are further classified by the appropriate season of their evening appearance: spring, summer, fall, and winter constellations.

There is no real starting place, since the constellations revolve, moving forward a "click" each night and returning to the same point each year. Depending on the time of night and which season it is, as well as which hemisphere you live in, a look at the night sky has the potential of providing you with a glimpse of the brightest planets, dozens of constellations, and thousands of stars and deep-sky objects. We'll start our tour with a look at the brightest stars as well as some of the best deep-sky Lawnchair-accessible targets, followed by an in-depth look at the ecliptic-traveling constellations — the zodiac — and the jewels that you can see there.

Star Names: Beetle Juice What?

The brightest stars have individual names originating from the native languages of the earliest Lawnchair Astronomers. Many are Greek, Latin, and Arabic names for mythological figures. The names Sirius, Canopus, Rigil Kent, Arcturus, and Vega are familiar to most astronomers as the five brightest stars. Even today, navigators of ships and planes call them by these names. But in order to more efficiently catalog the thousands of naked-eye stars in the sky, astronomers have incorporated a system in which each star is inventoried by its constellation, using the Greek alphabet as a name designation.

For example, Polaris, the North Star, or the polestar in the northern hemisphere, is the brightest star in the constellation Ursa Minor, the Little Bear. It is also known as alpha Ursae Minoris, or alpha *a* UMi as an abbreviation. The star Sirius is known as Alpha Canis Major or *a* CMa. Alpha Orion is the famous red giant, Betelgeuse. Also a tongue-twisting moniker if ever there was one! So, as the Lawnchair tour of the night skies continues, don't be confused when you sometimes encounter the use of two names for the same star. This is a key bit of Lawnchair Astronomy lore that will help you read the signposts of the heavens.

The Sheer Beauty of Stars

As we step out into the universe in anticipation of the awe-inspiring discoveries there — the galaxies, the star clusters, and the great nebulae — I'd like to point you to a subtle yet truly fascinating aspect of astronomy: looking at the stars — really looking at the stars. Though even I am often guilty of bypassing this activity and zooming right out into deep space, it is a completely worthwhile pursuit. Only by looking at and studying individual star systems, and the inherent beauty therein, will this become apparent to you too.

Accessible with binoculars as well as small telescopes, the bright spectacular stars, seen up close and personal, amaze even the most seasoned stargazers. The whole spectrum of colors isn't the only distinguishing feature you'll notice as you begin to familiarize yourself with the brightest stars in the sky. Some oscillate up and down in brightness, over periods as short as days and as long as months and years. Some of these variable stars have also served as excellent tools for measuring the universe.

Cepheid variables, named after the constellation Cepheus, where the first was discovered, are today so well studied and understood that they are looked for in distant galaxies to determine more accurate distance measurements. Their regular pulsations distinguish them, as a group, from other stars in the universe.

As mentioned earlier, many other stars are part of multiple star systems — stars revolving around each other or a common center of gravity. A long-term pursuit by many Lawnchair Astronomers is to resolve the separation between extremely close pairs. Often these pairs and triple systems are comprised of strikingly different stars, making the study of multiple systems even that much more mesmerizing.

The Brightest Star

Looking just at stars, and nothing else, can provide you with enough interesting sights to last a lifetime. For example, a popular component of winter's nighttime gallery of sky-paintings is Canis Major, the big doggie, which contains the sky's brightest stellar member, the Dog Star, Sirius. Easy to spot as it rises in the east during mid-winter, you should make it a point to get out and brush the snow off the Lawnchair and take a peek at this fireball. It is our brightest nighttime star not so much because of its intrinsic brightness, but because of its proximity to the Earth.

Sirius is the second closest visible star to us. It is the fifth closest overall — there are a few closer stars that are too dim to see without optical aids, the closest being Proxima Centauri. A great mystery surrounds Sirius though.

The earliest recorded information by notable astronomers such as Ptolemy catalog this hot, bright, bluish white star of today as being fiery red in color.

Several theories abound to explain these conflicting views including one by the author of Burnham's Celestial Handbooks. Burnham proposed that there might be a possibility that people in ancient times had a different aptitude for seeing colors. They saw colors differently than we do today, thus the reason for this anomaly. Theoretically, the human eye, which hadn't yet been exposed to "artificial" lighting may have recorded color and light frequencies differently from today's slightly more evolved, modern biooptics. Of course, this is a hard hypothesis to prove, but it is at least a possibility.

Another, more popular explanation exists. It was eventually discovered that Sirius was part of a binary system with a small, dim, white dwarf star orbiting extremely close to it. Scientists believe it is possible that the white dwarf may have been, as recently as two thousand years ago, what is known as a red giant star at the end of its reign. Once its fuel was depleted it settled

THE UNIVERSE: 1960'S TELEVISION STYLE

Our closest stellar neighbor is not Alpha Centauri, but Proxima Centauri, a dim and distant companion, and a component of a triple star system that was made famous by the sixties' television show *Lost in Space*, in which a family from Earth attempts to reach Alpha Centauri, "our nearest neighbor." Alpha Centauri, though brighter, is farther from Earth than Proxima Centauri (*proxima* comes from the Latin word for "closest"). A fifteenth-magnitude red dwarf star, Proxima Centauri can just barely be detected, and only in the largest Lawnchair light buckets or "LLLB's."

Alpha Centauri is another story though. It's the third brightest star in the entire sky and is easily found without the aid of any optical instruments. But, unfortunately for northern observers, only those living below the equator have this bright star rising on their horizon. Alpha Centauri is also known as Rigil Kent and has an almost equally bright, closely orbiting partner—a visual pair that can be split (observed as two distinct stars) with small telescopes.

down to its current state. So, if that was the case, its influence over Sirius when it was red was quite intense, causing those who studied this naked-eye star to believe it to be red. The question is not completely settled, and it may never be.

Take a look yourself, beginning late at night or early in the morning in the fall, or during the evenings of February and March as the Dog Star and Canis Major follow their master, Orion the Hunter, across the skies.

Pierre Laplace's Celestial Mechanics

Pierre Laplace (1749–1827) hypothesized the theory of the accretion disk, which today is still the standard model for the formation of the solar system. He was also instrumental, as were several other astute astronomers of his time, in developing the field of science we know today as celestial mechanics. Once the incredibly complex equations were worked out, without the help of computers, astronomers were finally able to accurately confirm and predict the positions and motions of the Sun, Moon and planets, a problem that had vexed those who studied the heavens since the first Lawnchairs were set up next to Stonehenge.

The Adventure Continues: Orion's Bright Red and Blue Stars

If you are the type who gets excited at the sight of really bright and colorful stars, then the constellation of Orion has just what the doctor ordered. Three bright blue sapphires comprise the hunter's belt, which is one of the well-known asterisms — recognizable groups of stars like the Big Dipper, the Pleiades, the Hyades V-shaped head of Taurus the Bull. Each of these three stars has a story to tell, but perhaps the most studied is the leftmost star of the belt, the one that fires or lights the gas and dust comprising the nebulosity that contrasts with the dark material of its famous neighbor, the Horsehead Nebula.

It is known as Zeta Orionis, or by its ancient name, Alnitak. In addition to its visual association with the belt stars, it is in its own right the primary member of a triple star system. Like its belt-comprising partners, Zeta is a very hot, young, bright star that is perhaps thirty-five thousand times as bright as our Sun.

Even though much excitement revolves around Zeta and its equine neighbor, its belt-mates, Epsilon Orionis (or Alnilam) in the center and Delta Orionis (or Mintaka) on the right, are each excellent targets for small instruments. They are each classified as supergiant stars and belong to multiple systems. Try comparing the three belt stars to each other.

Then, for an extremely contrasting view, take a peek at Betelguese, a considerably different sun reputed to be one of the very largest stars known. A red giant, Alpha Orionis (remember, many stars have two names) rests in the armpit of the hunter, and even with binoculars makes quite an impressive sight. Betelguese has my vote for the star most likely to go boom (super-

SUPERNOVAE

Supernovae are the most spectacular explosions ever witnessed by man. While normal novae are quite sensational, the sheer magnitude of the supernovae are astronomical in comparison. To attain supernova status, a star must first be extremely large, like Betelguese in Orion. Once past the point where all its available fuel has been exhausted, a gigantic star begins to feed on itself, burning heavier and heavier elements until the pressure and heat generated produces too much energy to be contained any longer and the star ends its life in one of the most cataclysmic events in the universe — a supernova explosion. These explosions are furious and leave nearly nothing where once resided an extraordinary sun. Regular novae occur quite frequently; on the average, thirty to forty per year in our galaxy alone, while supernovae are witnessed on the average of perhaps three or four per millennium (a thousand years). There is evidence that others have

nova) during our lifetime. If that ever happens — remember, you heard it here first!

Typically these types of stars, because of their sheer size, completely burn up all their available fuel within a few million years — as opposed to the conservative policies of our own five-billion-year-old solar furnace, whose projected supply of fuel will last perhaps another five billion years.

Another of Orion's more well-known members is Rigel, which also belongs to a multiple star system. Its partner, Rigel B, is itself suspect-

occurred in the Milky Way. Four have been recorded in relatively modern times.

A presently invisible member of Cassiopeia known as Tycho's Star was discovered in November 1572, and was the third of the four supernovae to be found within our galaxy. The earliest recorded supernova occurred in 1006 in the constellation Lupus. Next came the 1054 supernova in Taurus recorded by the Chinese, which is associated with one of the most famous novae remnants in the heavens — the Crab Nebula in Taurus. (When a supernova explodes, what remains is sometimes visible, as in the case of the Crab Nebula, as glowing nebulous material which astronomers refer to as remnants). Lastly, some thirty-two years after Tycho's Star was discovered, came the sighting of what is called Kepler's Star in Ophiuchus, the most recent supernova recorded within our own galaxy.

We have witnessed several supernovae in external galaxies, though, some of which were so bright that they outshone their entire home galaxies. A recent example was the supernova designated 1987a (for the year of its discovery) in the skies of the southern hemisphere, in a galaxy known as the Large Magellanic Cloud.

ed of being a close double star as well. But, because of their extremely minute physical separation (astronomically speaking), this can only be detected with the use of the most sophisticated professional devices.

In Orion, as throughout the entire night sky, the variety of star types and colors will astound anyone who takes the time to sit back, pour a cool one, and take the old Lawnchair out for an interstellar spin.

The Sky's Big, Lazy W

The constellation Cassiopeia, easily recognizable as a "Lazy W," is available in the backyard year-round, because it is one of the circumpolar constellations. Over the course of the year it appears to orbit around the North Star. Though not exactly centered on the true North Pole position, Polaris is relatively close to where that pivotal celestial coordinate is. From our vantage point (the Earth), over the course of an evening these constellations (Ursa Minor and Cassiopeia), as well as all the other pole-orbiting stars, circle this point in the sky. A camera centered on this point with its shutter open will produce spectacular star-trail photos. (See astrophotography section, page 42.)

Cassiopeia has Milky Way constituents, as well as external deep-sky phenomena — star clusters and galaxies. A comprehensive starmap will show where these wonderful treats lie. This duality is not unique to Cassiopeia and there are a few other constellations that are blessed with both local galactic star fields and extragalactic objects.

Cassiopeia is filled with sights well worth training a pair of binoculars or a small telescope on. Because it sits partially submerged in the Milky Way, a sweep across this celestial W will reveal thickly populated star fields and clusters, of which M-52 (one of Charles Messier's catalog objects) is one. This open cluster is comprised of hundreds of tiny stars mixed in with several closer and larger constituents. Containing both red and blue giants, this particular cluster is quite striking in both close-up and wide-field views.

The right arm of the W points up to M-52 (about the same distance away as the two pointing stars are from each other), and in the same field of view resides another peculiar inhabitant of Cassiopeia, the Bubble Nebula, classified as NGC-7635. (NGC stands for the New General Catalog, which was developed to record the thousands of

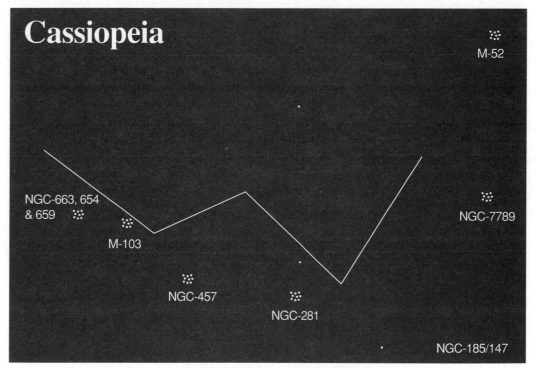

Cassiopeia

M-52

NGC-663, 654
& 659

M-103

NGC-457

NGC-281

NGC-7789

NGC-185/147

Many great sights await your inspection in Cassiopeia's domain of celestial jewels.

deep-sky objects in the night sky.) This nebulosity, which is only accessible through long-exposure photographs (see the section on astrophotography), has been classified as a planetary nebula, but speculation still exists as to its exact nature. Could it be the remnants of yet another giant exploding star?

M-103 is next on the Cassiopeian tour, the last object on Charles Messier's original list of 103 nebulosities. (Others have added seven more for a total of 110.) A small, loose cluster, it is a sight worth seeing with both small and medium instruments. You can find this very dim cluster near the bottom star of the left arm of the W.

Another fine Cassiopeian sight, especially in wide-angle views, is the triple cluster combination of NGC-654, NGC-663, and NGC-659, which lie a little farther up the left arm of the W, just above where you'll find M-103. Rounding off the

excellent clusters in Cassiopeia are NGC-7789, a dense cluster of small dim stars, and NGC-457, almost a twin of the double cluster in Perseus. This one is an easy target for Lawnchair instruments, lying the same distance as the triple cluster (NGC-654, NGC-663, and NGC-659) is from M-103, but in the opposite direction. Use the stars of the left arm of the W to point downward.

A pair of dwarf galaxies, siblings of the great Andromeda Galaxy, sit just within the border between Cassiopeia and Andromeda. NGC-185 and the more diffuse NGC-147 are both small satellites of the "great cloud" lying at a whopping 2.2 million light-years away. Under a dark sky many of these objects are attainable in Lawnchair telescopes. So are the massive Milky Way starfields in Cassiopeia, which even binoculars will reveal. This entire area of sky is splendid for touring the universe.

The Big Dipper: A Celestial Eye Exam

Most everyone in North America can point out a few of the most popular constellations: Orion, the Northern Cross of Cygnus, Leo the Lion, and so on. But none are as well known to millions as the Big Dipper, the asterism that is just part of the larger Ursae Majoris — Ursa Major, or the Big Bear constellation. Its companion, Ursa Minor (the Little Bear), also is recognized by its smaller, partially dimmer asterism, the Little Dipper.

A trick that every Boy Scout knows is how to use the Big and Little Dippers to find your way at night. By first locating the larger dipper of Ursa Major, trace from its handle to its "bucket stars." The outer two stars of the bucket form a pointer that is aimed almost directly at the North Star. Once you find the North Star, Polaris, it is easy to trace out the Little Dipper. Polaris is the end star of the Little Dipper's handle. By facing in the direction of Polaris, you are facing North, with

East on your right, South behind you, and to your left the western horizon.

As mentioned earlier, stars are often part of double and triple systems, revolving around a common center of gravity. A very famous pair found in Ursa Major are known as the Horse and Rider by Native Americans. Their ancient Arabic names, by which they are most commonly known, are Mizar and Alcor. As the middle stars of the Big Dipper's handle they are easy to find, and in ancient Greece were used as a test to see how good one's eyes were. This holds true today. Under decently dark skies, with any kind of luck, you should see what looks like a tiny companion close to, but definitely apart from, the primary component of what looks to be a binary system. What is additionally interesting is that these two stars are not associated with each other — each lies a substantial distance away from the other and us; but both just happen to lie along the same line of sight, from our perspective. Each, though, still belongs to a binary star system, having smaller orbiting companions, classifying them as double stars. Unfortunately, they are too close to split with Lawnchair instruments.

In addition to the bright main stars of the Big Dipper, the Ursa Major constellation consists of several more stars in what earliest Lawnchair Astronomers traced out in the imaginative shape of a bear. Within the borders of the Big Dipper lie several deep-sky objects. The galaxy M-82 (seen "edge on" from Earth) and its companion M-81 are just a few of the wonderful open-spiral type galaxies within the realm of the Great Bear. Easy year-round targets for Lawnchair instruments, they revolve with the Big Dipper, and never set below the horizon. A tour of this neighborhood is sure to inspire further trips back.

The Sombrero Galaxy in Virgo: Like M-82, this is a classic example of an "edge-on" galaxy.

THE DEEP-SKY WONDERS—LAWNCHAIR FAVORITES

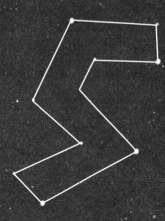

Orion and Winter Stargazing

Crisp and clear only begins to describe how truly rewarding the effects of winter's cold, dry atmosphere can be to a coal-black country sky. Even without optical aids, something special and unique seems to happen to the sky at this time of the year. The stars seem so close that you almost feel like you could fly up to touch them; so bright that you'd swear you could read by them. And if you use a pair of binoculars or a small telescope, the enjoyment factor can only escalate. This is what winter stargazing can promise. But I won't mislead you — even the staunchest Lawnchair Astronomer has had his doubts.

After a few minutes of stamping your feet to get the circulation back, you begin to have second thoughts. Maybe the old Lawnchair needs to be put away for the winter. *Not!* Though the thought has crossed many a northern Lawnchair Astronomer's mind, especially during the depths of winter when being outdoors at night means preparing for an arctic expedition! But once the sights of winter arrive, it gets a little hard not to go out and at least look for a minute....

Famous last words, or just a case of the dreaded astronomitus disease? Thoughts on a dark and moonless winter night can run the gamut. Am I crazy? What am I doing out in this subarctic environment, anyway? What could possibly compel me to endure these extreme conditions — of my own free will, at that? The answer is simple: the Orion Nebula.

The great nebula, M-42 in Orion, is perhaps one of the most spectacular objects in the entire sky. The area in and around the nebula is home to some of the most intriguing and thought-provoking sights in all of the heavens. Located in the middle of what is called the scabbard (the sheath swordsmen put their swords in), which hangs off mighty Orion's famous belt, is a seemingly dim though foggy star, just at the limit of naked-eye

RICHARD LOVISON

seeing. Up close via binoculars or telescope, you'll immediately see what all the hype is about. A longtime Lawnchair classic, this large and bright object has it all. You'll see bright stars and fuzzy clouds, and in the larger telescopes even the dark nebulae begin to stand out, back-lit and silhouetted by obvious stellar spotlights.

Four extremely hot, blue embryonic stars are what fire the nebula. They are known collectively as the Trapezium, with the main star of the multiple system called, simply, Theta Orionis. It has, perhaps, best been described as nanny to a set of newborn triplets, which on a good clear and dark night you may glimpse in any medium-size instrument.

The area out from the central core of the Orion Nebula is made up of various twisting and turning wisps of gaseous and dusty material. The darker matter visible in very large instruments is also quite vivid. You'll notice a greenish hue, which is caused by the ionization of oxygen molecules prevalent throughout the cloud. In deep-sky

Happy Hunting Ground: The Orion Nebula (M-42)

photographs blues, reds, and purples are also highlighted, though are not seen visually because of the low light levels. In wide-angle views, the nebula is accompanied by a small nebulous patch enveloping a few stars to the north and Iota Orionis to the south.

Another, perhaps not as famous, sight in Orion, though accessible with binoculars or a small scope, is M-78. Just above the left star of Orion's belt, Zeta Orion, this bright section of the nebulosity that actually covers most of the Orion constellation is interesting to trace out. And the tour wouldn't be complete without mentioning an especially surreal sight near Zeta Orion.

In a shared wisp of this same nebulous material sits one of the most celebrated subjects of telescope advertisements: the coal-black Horsehead Nebula. A great example of dark nebulae, this fortunately situated object, spectacular in photos, is a difficult Lawnchair target, but one you'll be glad to add to your deep-sky Lawnchair collection. Don't expect to see in your eyepiece what only long-exposure photos enlarged ten times over reveal! What you will see in the largest backyard instruments, at most, is a very small, dark blemish, which is barely discernable as the horsehead, protruding out into an otherwise bright, gaseous nebulosity. This one could earn you the Lawnchair Eagle-eye Merit Badge.

The Other Little Dipper

In early winter the constellation of the bull, Taurus, rises in the East and is easily recognized by its elongated V-shaped face. Placed close by, the asterism called the Pleiades (play-ah-deez), or Seven Sisters is part of an age old mystery: only six sisters seem to be left these days. During ancient times, Lawnchair Astronomers noted that while looking at this tiny group of stars, which by the way is often confused by the uninitiated as the Little Dipper, seven stars were visible via eyes alone.

Today, without the aid of binoculars or a telescope, a quick look seems to reveal only six visible stars. However, many

The Pleiades (M-45): How many sisters?

Lawnchair Astronomers have reported seeing up to fifteen and sixteen stars with eyes alone. It's

RICHARD LOVISON

been reported that as many as twenty are possible under ideal dark-sky conditions. With the aid of a telescope, though, this is one of the most fabulous sights in the sky. Classified as an open cluster (a loose conglomeration of stars, as opposed to a globular cluster, where stars are seemingly globbed together), the six sisters become hundreds of blue-sapphire bright jewels.

A nebulosity surrounding the Pleiades is faintly discernible from very dark-sky sites, though in deep-sky, long-exposure photographs, the majesty of this seemingly brushed-on fog is absolutely striking.

From late nights in August and September, when it rises in the East, to the early evenings in March and April, when it chases the Sun down to the western horizon, the Pleiades makes for an excellent backyard target. And this is one of the easiest objects to find and show to your many gathered and adoring Lawnchair admirers.

PLEIADES

At a time in our distant past, before the advent of space-based instruments and 200-inch telescopes (or larger), before any telescopes were invented, for that matter, some of our ancestors were quite prolific, never mind extremely imaginative, in the telling of stories about the stars in the sky. Many stories revolved around what we know as the Pleiades, or the Seven Sisters.

One story in particular had the sisters being protected from Orion the Hunter, who chases them across the sky each year during the winter months. Still other stories involved earthbound virgins being placed in the sky for safekeeping. A modern addition to the myth has one sister married to a mere mortal, while her siblings' nuptial encounters were with gods exclusively. For her apparently shameful indiscretion, the seventh sister was hidden away from the world in shame, or so the story goes.

The Great Little Cloud

Located just outside the great square of the constellation Pegasus and doubling as the tail of the winged horse, is a portion of the constellation named for princess Andromeda. When you look at the large, diamond-oriented box of Pegasus rising almost directly overhead as you face east in late summer, the leftmost star of the square, Alpheratz (or Sirrah), is the alpha, or primary, star of the constellation Andromeda. Envisioned as lying on her back in a reclined pose, the mythological princess contains the

most distant object (2.2 million light-years) in the entire sky that can be detected with eyes alone.

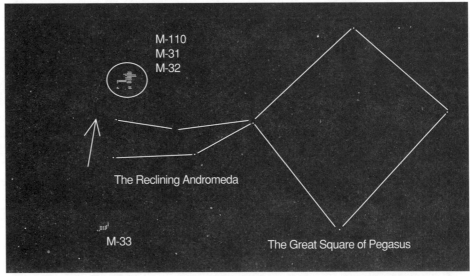

Find the Square of Pegasus, then step-stone your way along the Andromeda Constellation. Use the second set of stars from the square as pointers to the Andromeda Galaxies.

There are several distant objects that we can see within our own galaxy including all the stars, varied star clusters, and nebulous-type objects. None of them compare, though, to the enormous, and often photographed, M-31: the Andromeda Galaxy. M-31 is located approximately where the princess's right (upper) shoulder or elbow would be. Visible with eyes alone, you will have to use averted vision (looking slightly away from an object to let the more light-sensitive portions of your eye focus on the intended object) to spot the seemingly flickering image at the threshold of the eye's limiting magnitude level. It can be done easily, especially in a clear, dark sky.

RICHARD LOVISON

Andromeda (eye) strain? At 2.2 million light-years away, M-31 is the farthest night-sky object visible to the naked eye.

With binoculars, the cloud is easily discernable as an oval patch, which appears to brighten at the center. In small to medium-size telescopes the companion ellipticals — small oval satellite galaxies, are visible as well. Directly across from the princess's famous constituent, at her lower or left elbow, lies another galaxy that is perhaps an even more spectacular sight in large instruments than its famous sibling.

Using the "giant chicken" method described on page 77, begin in Andromeda, then sweep down across the sky where you will come across a large, dim galaxy known as M-33, or the great Pinwheel Galaxy. The pinwheel actually resides in the constellation Triangulum. An almost face-on spiral, in deep-sky photos the knotty arms are quite apparent and visually appealing. Viewing M-33 with field glasses is possible under very dark skies. Although it is almost the size of the Moon, many stargazers may miss this large but dim haze unless they are very careful. It does have a reputation for being quite elusive. Do not try to overmagnify; a wide-field lens will serve you best in the search. Don't be disappointed if you don't see what is visible in the long-exposure photographs.

The Great Cloud, M-31, is the Andromeda

ANDROMEDA AND OUR OTHER GALACTIC NEIGHBORS

Thirteen thousand quadrillion miles — give or take a few billion — is how far Andromeda is from us. It is the closest and perhaps the largest of all the spirals in the local group. (The Milky Way is considered a twin to the Andromeda Galaxy, and each belong to a collection of galaxies that are seemingly traveling together through the universe at large, perhaps even belonging to the larger Coma-Virgo galaxy cluster — see illustration on page 128 Libra/Virgo/Leo.)

The only visible galaxies closer to the Milky Way are the Large and Small Magellanic

constellation's main attraction, and much has been written and learned about this island universe, perhaps the mirror image of our own galaxy, the Milky Way. This naked-eye object was wondered about as far back as A.D. 905 and was even known as the Little Cloud. It could be found on early star maps even before telescopes were invented. In the seventeenth century, Simon Marius was credited as being the first to telescopically peruse the elongated, dim patch of light. For some time it was theorized that M-31 was a forming solar system, and it wasn't until spectroscopic analysis (a process used to examine the light spectrum of an image) that its nature was confirmed and it was finally proven that this object's light originated from numerous individual stars.

Clouds, which are irregular galaxies, just satellites to the Milky Way. (They are viewable only from extreme southern hemisphere locations.) In addition to these, what was previously thought to be the very closest galaxy to the Milky Way, a small dwarf galaxy known as Snickers — which had been detected only as a radio source in the direction of the red giant Betelguese, in Orion, has recently lost that unique status.

In the spring of 1994, another irregular-shaped dwarf galaxy was detected peeking up over the top of the central hub of the Milky Way, perhaps in the process of merging with our home island. If this is the case, it would be the closest galaxy to ours.

Seeing the Andromeda Galaxy will make a great conquest for your "what I've seen with my very own eyes" log. The well-versed Lawnchair Astronomer will make this a primary stop for lingering late-summer Lawnchair admirers. How many people can claim to have seen in person something that is 2.2 million light-years away?

Hercules

In the summer, if you look to the East in the late evenings, you can see Hercules rising off the horizon, just below the northeast corner of the easily identifiable crown-shaped Corona Borealis constellation. An odd, box-shaped structure that reminds me a little of Orion, minus the belt, the ancients saw it as the great mythological warrior Hercules, and he can be found just a stone's throw from the curving crown shape of Corona Borealis.

On the right side of what you might call the upper box of this constellation is what has been described as the most spectacular globular cluster in the sky. M-13 was described by Charles Messier as a "round nebula containing no stars." Of course, this was directly related to the crudeness of his instrument. Today's Lawnchair-grade telescopes resolve this cluster easily into the thousands of stars that make it up.

Another fabulous cluster found in the Hercules constellation is known as M-92. It, too, was first found in the eighteenth century by Charles Messier. In wide angle, these clusters present themselves as bright patches of light with perhaps a smattering of resolved stars near their edges. With medium-size instruments, each cluster can be magnified to the limit of the available optics, where further resolution is not possible. In 12-inch and larger instruments, you can expect to resolve much more of the thousands of stars here, which in smaller instruments are impossible to discern. Plan a stop here during your next Lawnchair tour of the summer skies, though that may present you with a slight problem: trying to get a turn at the eyepiece!

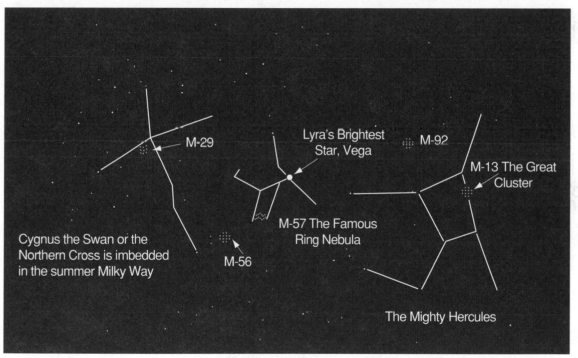

M-29

Lyra's Brightest
Star, Vega

M-92

M-13 The Great
Cluster

M-57 The Famous
Ring Nebula

Cygnus the Swan or the
Northern Cross is imbedded
in the summer Milky Way

M-56

The Mighty Hercules

*Some of the summer's most fascinating objects can be found in this area of the sky. Overhead, use the large cross of
Cygnus the Swan to find Lyra and the Mighty Hercules.*

Lyra, Vega, and the Ring Nebula

You will have no trouble spotting Vega, the Falling Eagle or Harp Star. Move east — northeast from the large box shapes just overhead, of the mighty Hercules to the top of the K-shaped constellation, Lyra. Vega is also the brightest star of the Summer Triangle, joined to Deneb in Cygnus and Altair in Aquila. It is also the fifth brightest star in the sky overall. Once you train any type of optical aide upon this fiery spot, you'll at once recognize the immensity and grandeur of this star. With eyes alone it can easily be taken as just another bright star. This is quite deceiving, as you'll discover upon a closer look.

Twenty-seven light-years distant, Vega's actual luminosity is reported to be 58 times that of our pale yellow Sun. Described as cold blue, it is twice as hot as our Sun and three times as massive. This star's radial velocity (its own movement through space) will place it, from our perspective twelve thousand years from now, in the position of our celestial polestar, which, of course, today is filled

by Polaris. This was where Vega was originally, some twelve thousand years ago. Another peculiar aspect of Vega's position is that it is where we are going — that is to say, our Sun's direction through space is toward Vega and the Northern Cross. Don't start packing just yet, though. It will take a leisurely 450,000 years to get there.

The Lyra constellation, named after any of several lyre-carrying celestial messengers, is a rather small group of stars headed by the bright Vega. Lyra sort of resembles the letter K, and it is easy to find between Hercules and Cygnus — the Northern Cross — all of them high in the sky late nights during the summer. The special treat here, of course, is the extremely famous and often photographed Ring Nebula, M-57.

This tiny "smoke ring" is known as a planetary nebula — a ring of ejected material originating from an energetic parent star — but it has no real relation to planets. A mystery still exists as to the exact nature or cause of these types of gas

A smoke ring in the sky: The Ring Nebula (M-57) in Lyra

clouds. Some feel that they are remnants of ancient novae, and some feel that they may in fact be the precursors to planetary development. Those nebulae containing central stars, as does the Ring Nebula here in Lyra, are the strongest candidates for planetary system-hood.

Found easily in telescopes of three inches or more, the substantially larger backyard instru-ments will produce a much more dis-cernable view of this most famous of all "planetaries." Located between the two bottom stars in the Lyra constellation, the Ring Nebula should provide you with an achievable Lawnchair challenge: seeing the sole star that sits, waiting to wink back at you, from the center of the ring. Local weather and light conditions will play a major factor in your success.

By mid-summer, following Hercules in the sky in late evenings, you'll find the area of Lyra quite dark, as it is almost overhead. At the zenith — the point directly overhead — view-ing conditions are usually best, as this is the very darkest part of the sky, as well as the portion of sky where you are looking through the least amount of obscuring and sometimes even opaque atmosphere. Once you capture this target for the first time, it will forever be part of your must-see list of summer's Lawnchair jewels.

Cygnus and the Milky Way

Next on our celestial journey is Cygnus, home to numerous fine sights, many of which are contained in the massive galactic star fields that inhabit the region. We can't talk about Cygnus without mentioning its most famous constituent, the remnant of a thousand-year-old supernova, first witnessed by the Chinese in the eleventh century. The object Cygnus X-1, in addition to its visual counterpart, an unimpressive 9th-magnitude star, is an impressive X-ray source and was one of the first candidates for proof of the existence of black holes. Because of the nature of black holes (a black hole is an incredibly massive object that has been so compressed, it develops a gravity that is so strong, not even light can escape its clutches), they cannot be detected directly. Astrophysicists have theorized for decades on how a black hole might be detected. One theory states that evidence of a black hole might be detected as a source of energy eminating from the X-ray portion of the electromagnetic spectrum. This is what has been observed in and around the area of Cygnus X-1.

Among the other fascinating sights in Cygnus, at the top is its primary star, Deneb, one of the greatest supergiant stars known. Its actual luminosity has been calculated to be in excess of 60,000 times that of our Sun. At a distance of sixteen hundred light-years, our Sun seen at this distance would be a telescopic target only, at a dim 13.3 magnitude. Unlike Sirius, which owes its local brightness to its relatively close proximity to us, Deneb shines at above first magnitude because of sheer size, about sixty times that of Sol.

The most fun you can have in Cygnus with binoculars and small telescopes is to tour the Milky Way along the longer axis of the star-embedded Northern Cross, which our Lawnchair ancestors saw as a Swan. This portion of the sky is bursting with massive pinpoint star fields. Rising up from the East Northeast every summer, the Northern Cross stays with us all summer long, its foot pointing south along the path of the Milky Way.

Looking at Cygnus from a dark-sky site, you can see how the Milky Way got its name. This band of light shines from the sheer intensity of the cumulative effect of innumerable gatherings of stars. From Cygnus to the southern horizon, the summer Milky Way can provide a lifetime's worth of discoveries. Saturated with a variety of Messier and NGC objects, the Milky Way is the proverbial Lawnchair treasure chest of the sky.

GREAT MOMENTS IN LAWNCHAIR ASTRONOMY

Halley's Other Feat

During the eighteenth century, Edmund Halley (1656–1742), besides discovering the famous comet bearing his name, was responsible for determining that the stars were not fixed objects but did indeed have their own motion, known as Proper Motion, detectable over a period as short as just a few years.

Ophiuchus: Bright and Dark Clouds

Our next stop is Ophiuchus, a giant house-shaped constellation just below Hercules. Home to the center of the galaxy, Ophiuchus is filled with astronomical treats. First off, it contains the largest number of globular clusters of any single constellation. Also imbedded in the summer Milky Way are Sagittarius and Scorpius, in second and third place respectively when it comes to globular clusters.

Ophiuchus is also home to Barnard's Star, the fastest and second closest star from our Sun. Its incoming velocity will place it as the nearest star to the Earth, surpassing the Alpha – Proxima Centauri system, in about eight thousand years. By then it will be a mere three and a half light-years away.

The section of Milky Way visible in Ophiuchus is the proverbial pièce de résistance, and contains some of the most curious sights in the galaxy. Black nebulae (or "the dark clouds") show up as seemingly three-dimensional images in the eye-piece. Silhouetted in the foreground, they block out the light of the thousand pinpoints that rest behind them from our point of view. Some of these clouds, which are made of both light and dark matter, dust particles and gas, are larger than our entire solar system. As you scan across Ophiuchus, use a wide-angle eyepiece to take in as much of the massive star fields as you can. With larger instruments you're sure to spot several of the aforementioned dark clouds.

Looking up at Ophiuchus, embedded in the summer Milky Way, I can't help but visualize it as a giant celestial barn — a wonderful place for the Lawnchair night-owl set to hang out!

THE ZODIAC

A Year's Worth of Ooohs and Aaahs!

From both northern and southern hemispheres, the ecliptic marks the path that the Sun, Moon, and planets travel across our sky. As the sky passes by month after month, in addition to the Sun and planets, new constellations rise in the East while those that have been up for several months begin to set in the West. Those star pictures that travel along the ecliptic are known as the zodiacal constellations, or the zodiac.

Their names are almost all based on mythological figures, given to them by the Lawnchair Astronomers of antiquity. Both animals and consorts of the gods are represented in the twelve constellations of the zodiac. What we're after, though, are some of the best deep-sky objects, which are embedded in these twelve sky murals. Our tour will concentrate on the most spectacular of these, those that are known as the cream of the Lawnchair crop! The tour will begin with Aquarius, which rises in the late evenings of July and remains in the sky until January, when it floats above the western horizon by nightfall.

Aquarius: The Rainy Season

One of the zodiac's larger constellations, Aquarius the Water Bearer, is supposed to look like a figure with a vase or ladle, the latter of which is suspended above the downward-pointing arrowhead shape of Capricorn. In some parts of the world, especially during ancient times, the rising of Aquarius signaled the oncoming monsoon seasons and the arrival of Autumn as well as Winter's constellations. In modern times, the Age of Aquarius was immortalized during the sixties by the Fifth Dimension.

And there is reason to sing about Aquarius and what there is to behold in her arms. Even with small instruments, M-2 is visible as a small, round, fuzzy star. And in larger, 8- to 10-inch telescopes, this distant cluster will partially resolve into its myriad stars. Only in the largest amateur and professional telescopes will more detail than this be apparent. Other star clusters in Aquarius include the spectacular globular, M-72, and the asterism known as M-73, a loose gathering of a few close stars that seem to draw the eye. M-72 in extreme western Aquarius lies some three degrees West Southwest of NGC-7009, the Saturn Nebula.

In 1782 William Herschel, while searching through Aquarius, came upon this very bright and extraordinary planetary nebula he named, appropriately, the Saturn Nebula. With small telescopes, not much more than a small, dim cloud will be apparent, impossible to see at all under anything but optimum conditions. With 10-inch or larger reflectors, the ansae (ring-like protrusions) are visible under dark conditions. With larger instruments, the object itself has been described as striking, with a vivid green glow caused by strong ultraviolet radiation coming from the central star, which shines at a warm fifty-five thousand degrees Kelvin. Distances to planetary nebulae are often guesstimates, and this one works

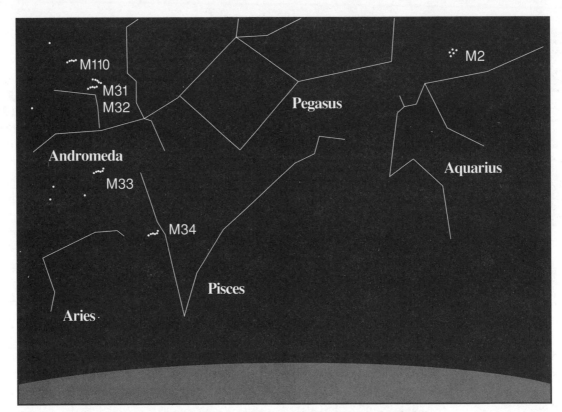

The rainy season?: Aquarius and the other constellations of late summer and early fall

out at approximately four thousand light-years distance, give or take a hundred or so. The problem is partially complicated due to conflicting measurements between the central stars of these types of nebulae and the nebulae themselves.

Another Aquarian nebula, NGC-7293, is one of the largest and nearest of the group known as planetary nebulae. Visually it presents itself as half the size of a full Moon, though this ring-shaped object is quite dim, having a very low surface brightness. Accessible with binoculars under perfect seeing conditions as a dim, hazy ring, with small telescopes the Helical Nebula will be best viewed with a wide-angle eyepiece. The best instrument with which to view this target would be any of the "rich field" telescopes (those with wide fields of view) with wide-angle lenses added. In photographs taken with the Palomar 200-inch telescope, a wealth of detail is revealed in this nebula.

These planetaries are peculiar objects, to say the least. There are over five hundred known examples, though ten thousand may exist in our galaxy alone. Planetary nebulae get their name because of their resemblence to a planetary system, though they have nothing in common with planets. Typically, they are recognized as energetic shells of gas (the Ring Nebula in Lyra is an excellent example) surrounding a star that is going through some dramatic, though sometimes slow, changes. One theory proposes that they may possibly be caused by lazy or slow novae processes, instead of the explosively energetic and typically short-lived supernovae. Often difficult to capture in Lawnchair instruments, planetary nebulae are some of the most sought-after trophies. Almost everyone's seen Saturn, but who's seen the Saturn Nebula?

Pisces: Something Fishy in the Sky

Following Aquarius in the late evenings of August, Pisces lifts off the eastern horizon chasing the great square of Pegasus up into the sky. Pisces is quite easy to spot as the giant letter V whose right arm ends in a small circlet of stars. As a celestial puzzle piece, Pisces almost looks as if it fits up against the eastern corner of the big diamond-oriented box shape of Pegasus. Of course, Pisces is supposed to represent two fish swimming the heavens, but it's easier just to look for the V asterism.

We start our tour with Alpha Piscium, at the bottom of the V, or as the Arabs called it, "the cord." Its Arabic name, Al Rischa (or Okda), refers to the flaxen cord that was said to have been used for binding the two fish together. I understand that they also used this same flaxen cord to secure their Lawnchairs up on their camels, which took them to and from their favorite desert dark-sky sites. Alpha Piscium,

a double star just made for Lawnchair instruments, has been the subject of controversy for ages: reports of this pair's color have gone from greenish-white and blue to pale or brown yellow. One early twentieth-century Lawnchair Astronomer, who was unwilling to be identified, offered: "Just a pair of weird-colored stars" — apparently all one dared say in the late twenties. Suspected of being a variable star (a star whose brightness changes over time), the primary of the group has been studied at length, and still no precise determination has been made in regard to this Piscean constituent. Find and take a close look at this star pair, and see what color you think it is.

Another interesting star in Pisces is called Van Maanen's Star, located a few degrees south of Delta Piscium. This is one of the very few white dwarf stars that can be seen with Lawnchair-grade instruments. One of the smallest stars known, it is also possibly the closest white dwarf star to our solar system, except for a few faint companion stars to Sirius and Procyon. Although it is almost as small as Earth, it has the approximate mass of our Sun, which makes this an extraordinarily dense twinkle! It's also a member of a group of stars considered the oldest in the galaxy. Today it's basically a dead star: a mere 13.8 light-years away, this star, which is no longer producing any real energy, is like a morning-after log in the fireplace, which has been reduced to so much gray ash and a few dimly glowing orange embers.

Next up the right side of the V is Eta Piscium, and a hop, skip, and a jump to the East Northeast from there lies M-74. A perfect example of a face-on galaxy, this Messier object was first seen by one Pierre Mechain, who described it as a nebula that contained no star. A month later, old Chuckie Messier added it to his now-familiar list. As time marched on, and larger, more powerful instruments became available, it was successfully photographed, and all doubt as to its true nature was resolved. What we believe our own galaxy would look like if seen from this angle, M-74 is classic in its shape and form, also resembling the Great Spiral (M-101) in Ursa Major.

Also within the borders of the zodiacal fish lie a few peculiar galaxies, one of which seems to be going through some traumatic changes and upheavals. NGC-520, as seen in photographs taken through the giant 200-inch telescope at the Palomar Observatory, may perhaps be the result of a pair of colliding irregular galaxies. Finally, the galaxy NGC-128 has a peculiar box shape to it, making it quite unique in all the heavens. Both of these are only open to view in deep-sky photographs, but they're still quite interesting nonetheless.

Aries: Skip This Part

Late in September, if you face southeast and look along the left side of Piscean V, you'll see a grouping of stars that your earliest Lawnchair brethren thought looked like a sheep. Exactly what it was that they were drinking or smoking in those days still hasn't been determined official-ly, but suffice it to say, "Thank goodness they were only driving Lawnchairs!"

Generally speaking, though, Aries is a fairly boring piece of sky. Its alpha star, Hamal, "Head of the Sheep" is only a magnitude-two star. In deep-sky photographs, a handful of 12th magni-tude galaxies pepper the area, and other than a few double and triple star systems, you could safely say that Aries won't be found on too many Lawnchair showcase lists.

There is one note of trivia, though, that any Lawnchair Astronomer worth his or her salt should know. The saying "In like a lion, out like a lamb," though often thought to relate to the weather in March, actually comes from the skies. The constellation of Leo the Lion is seen to be rising in the East at the beginning of March, while at month's end, Aries, or the Lamb, is setting in the West. Coincidence? I dare say not! Just where is Oliver Stone when you need him most?

Taurus

If you're out to see Taurus as it rises in the East during the crisp, cool evenings of October, look to the East for another V-shaped asterism, but this one will be oriented on its side. This is the face of the bull. As winter encroaches, so does the bull, rising high up along the autumn ecliptic.

It also the time of year that harbors a change in the weather, often evidenced by the thin morning frost you see on the ol' pumpkin as you pack up your Lawnchair and head in for a good morning's sleep.

Halloween, the holiday of dark things, signifies for Lawnchair Astronomers the return of the stellar giants, the winter manifestations of the hunter and the twins, the bull and the seven sisters.

In addition to the aforementioned Pleiades, the constellation of Taurus the Bull contains many celebrated Lawnchair targets. One such widely popular treasure is the Hyades star cluster. This large, open cluster, along the left side of the triangle-shaped face of the bull, presents itself in either small telescopes or binoculars as a wonderful jewelry box of sparkling, bright stars. To the naked eye a hint of what is there is all that is possible.

Another famous sight found on the extreme northeastern outskirts of Taurus is the first of the Messier objects, M-1, the Crab Nebula. As mentioned earlier, supernovae are relatively rare and only four have been recorded to date. The Crab Nebula rests just northwest of Zeta Tauri, which marks the tip of the bull's horn, and is the remains of one of the most famous of these explosive events. At 8th magnitude today, this object is absolutely possible to see in Lawnchair grade instruments — though the larger the better, as this is a faint and wispy nebula.

In the radio spectrum — radio telescopes can detect many signals or objects that have and sometimes don't have visual counterparts —

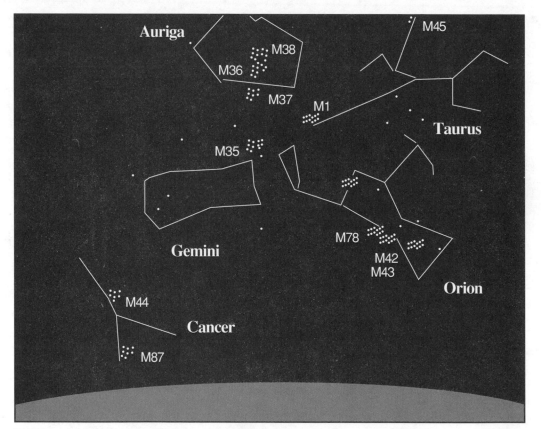

Bullish on winter astronomy: Taurus and the constellations of late fall and winter

Crab Nebula (M-1): Remains of a supernova

this is one of the four brightest sources in the heavens. In addition to the nebula, something else lives here. A highly energetic radio source pulsating in regular, high-speed increments, the Crab Pulsar is thought to be the detectable result of a neutron star, or a star that was per-

haps just not quite big enough to end up as a black hole when its density became great enough at the end of its life to bring on its collapse. The Crab Nebula and Pulsar is what remains of a star that was perhaps similar to the red giant Betelguese.

At a minimum of at least several solar masses (the size and weight of our Sun) after the explosion, though not capable of black-hole, light-eating status, the core of a star that was once perhaps as large as our entire solar system had been reduced to an incredibly dense rotating sphere occupying a space smaller than the Earth's Moon.

Also astonishing is that the M-1 Pulsar blinks in both the radio and X-ray spectrums. The cause of this pulsating is attributed to the core's incredible rotational speed. Similar in operation to a lighthouse beacon, as these objects rotate they emit powerful beams of energy, which we see as pulses (as fast as thirty pulses per second) in both the radio — and in the case of M-1 — in the X-ray portion of the spectrum.

When NASA's SETI (Search for Extraterrestrial Intelligence) astronomers first detected these signals, many thought that they just might be coming from an intelligent extraterrestrial source — perhaps they were a type of navigational space beacon for a space-faring civilization. Once all the facts were in, though, the theory that we know today as the neutron star was confirmed and realized again and again throughout ours and other nearby galaxies. The Crab Nebula and Pulsar is only one of hundreds of these objects catalogued so far, though none has been studied as extensively as the M-1.

Lastly, another intriguing aspect of the Crab Pulsar is that it seems to be variable in its pulses, and the reasons for this still have to be explained. But many leaps forward have been made in the past several decades in our understanding of the processes resulting from dying and exploding stars. We continue to learn more every year as new experiments are devised to study the heavens.

Taurus indeed makes for a great place to start an autumn Lawnchair tour of the zodiac. In good company, the Twins Gemini follow as the night and the year progresses.

Gemini: Sons of Zeus

The stick figures belonging to the Gemini Twins are quite apparent in the Eastern skies by late evening in November. Gemini proves to be a grand hunting ground for bright and distinctive stars. The head stars of the Gemini twins, for example, have been the center of attention for millennia. Depicted on the earliest coins and other varied ancient artifacts, Gemini's two brightest members, Pollux and Castor, are the eyes of the fabled twin sons of Zeus.

Upon closer telescopic examination, Castor, the dimmer of the two, has some surprises waiting for you. Not just one or two, either: Castor contains three stars in close proximity to each other. What's more fantastic than that is that each of the three are members of even closer binary star systems. So where you see with your eyes alone a single bright star, in actuality resides six separate stars.

Gemini also contains a few excellent deep-sky treats. First and foremost is M-35, a thickly settled star cluster near the foot of the right twin that has been admired by Lawnchair Astronomers for centuries. In its immediate vicinity also lives another much more tightly packed cluster cataloged as NGC-2158. In binoculars, M-35 may be resolvable under perfect seeing conditions, though NGC-2158 will at best seem a blur of nebulosity. In a good-size instrument, though, both are wonderfully bright and exciting sights.

There are other clusters in Gemini that Charlie Messier seemed to have missed. NGC-2266 and 2420 are both fine telescopic sights, but just haven't got the media attention that their sibling M-35 has. Maybe if they just went on *Hard Copy* ... I wonder if there are any openings for Lawnchair paparazzi!

Cancer: Star Clusters for Every Season

Creeping along the ecliptic close behind the twins is the constellation of the Crab. Cancer, usually identified as the inverted letter Y, might have had a different name had it been left to modern Lawnchair Astronomers. Just imagine the fun when someone asked you what your sign was and you could say, "I'm a Lobster!" Home to some truly awe-inspiring stellar congregations, Cancer is quite an interesting constellation to learn.

Just above and to the right of the central star in the inverted Y of Cancer lies M-44, the Praesepe or Beehive Cluster. Resembling several fiery bees buzzing around the nest, once you see this for yourself you'll know why it was named as it was. It is quite easy to find and visually appealing even with a pair of binoculars. It is best seen in a wide-angle or a rich-field telescope lens. Known as a loose or open cluster, it comprises over two hundred stars. With his crude instruments, Galileo counted a mere thirty-five stars. Before his time, this cluster, visible to the naked eye as a small nebulosity, was used to predict weather patterns. If it was not visible, then it was assumed that stormy or cloudy skies were imminent.

Slightly below the left foot of the Cancerian inverted Y resides another, and even more spectacular star cluster, M-67. This conglomeration counts well upward of five hundred stars between 10th and 17th magnitudes, which warrants at least a mid-size 8- to 10-inch telescope to resolve it into its constituent members. Though it appears nebulous at best in binoculars, a moderate-size telescope will resolve the group. A moonless and clear night is a prerequisite for capturing this Lawnchair jewel.

Leo: Bright Spots on the Big Cat

Leo the Lion can be found high overhead by spring. It ushers in warming temperatures and an area of sky that looks out onto the universe filled with hundreds and thousands of galaxies.

Famous enough in its own right as a longtime Lawnchair stop, Leo has a number of interesting components. First off, at the lion's front shoulder lies its primary star, Alpha or Cor Leonis, also known as Regulus, the little King. Quite appropriate, since we're speaking about the king of the jungle, Regulus was regarded by the ancient Persians as one of the four Royal Stars, brethren to Aldebaran, Fomalhaut, and Antares.

Though unaccessible by Lawnchair instruments, another point of interest within the boundaries of the Leo constellation is a star known simply as Wolf 359. Named after the astronomer who first identified it from photographic plates, it was confirmed that this particularly dim star, a small red dwarf, is the third closest star to us after the Centauri group and Barnard's star in Ophiuchus.

The first of the Messier objects we'll find in this area of the sky are a pair of very fine galaxies, M-65 and M-66, each classified as spirals. Easily viewed through binoculars and small scopes under dark skies, they lie in a line just below the back paws of Leo. Easily seen in the same wide-field, telescopic view, these two are accompanied by an edge-on galaxy known as NGC-3628. It is thought that perhaps these two island universes are part of the larger and well-populated Coma-Virgo cluster of galaxies.

A popular Lawnchair pursuit is to try to locate these three galaxies and note the differences in their attitude toward us and each other. These three galaxies are well worth the hunt and are great prizes to add to your list of targeted and found Messier objects.

Springing ahead with the big cat: Leo and the constellations of spring and early summer

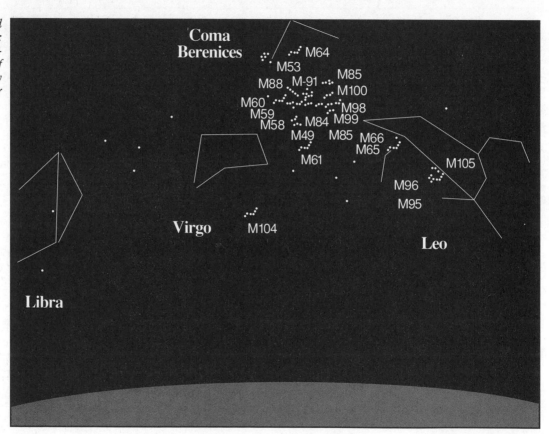

Coma Berenices

M64
M53
M85
M88 M-91
M100
M60 M98
M59 M99
M58 M84
M49 M85 M66
M61 M65
M105
M96
M95

Virgo M104

Leo

Libra

Virgo: Where Galaxies Play

Following in the paw prints of the most regal member of the zodiac comes none other than the constellation Virgo. Virgo, or the Virgin, is home to Spica, the star representing the ear of wheat the virgin holds in her left hand. Spica is the perfect example of a 1st-magnitude star. Although it varies just above and below the magnitude-one classification, it is listed as the 16th brightest star in the sky.

A spot between the body of this celestial maiden and the lion's tail is where most Lawnchair instruments end up pointed after a peek at the virgin. The Coma-Virgo cluster resides in this little corner of the sky. Not a star cluster, mind you, but a cluster of galaxies. When old Charles Messier first cataloged these little light clouds in the late eighteenth century, he had no idea of what they were.

In this century we have learned that we are circling our own spiral-shaped star city in what is known is the Orion arm of the galaxy. We've additionally learned that our Milky Way is but one of hundreds of thousands of galaxies populating the universe. Virgo is one of the only galaxy clusters that can be detected in lawnchair-quality (and -sized) instruments. There are many groups of galaxies that lie even farther away, which only the largest instruments can detect.

There are over three thousand galaxies peppered throughout the immediate vicinity of the Virgo "cloud," many of which are only detectable with the use of astrophotography and long-exposure photographs. A hundred of these are visible with a good 8-inch telescope from a truly dark-sky site. A dozen or so are able to be viewed with smaller instruments. The Virgo cluster is the nearest of any of the galactic clusters, after what is known as our "local group," which itself contains twenty or so members. It is thought that our local neighborhood is just part of the larger Coma-Virgo group.

The galaxy M-51, as seen through the powerful telescope at the U.S. Naval Observatory

Scanning this area of the sky with your telescope will reveal a large number of small, wispy light-clouds. Here you'll see what Charles Messier saw in his crude telescope. He named a dozen or so of these tiny clouds, M-49, M-51, M-60, M-84, M-85, M-87, M-88, M-90, M-91, M-98, M-99, and M-100. These were visible to him two hundred years ago. Which ones can you see today?

On the Coma-Virgo border lies some of the brightest members of these "star cities": M-84 and M-85. The exact distance to the cluster is still disputed, but it centers between 42 and 70 million light-years away. So remember as you gaze at this area of the sky that you are not only looking at galaxies that are 50 million light-years away, but, at galaxies as they were fifty million years ago.

Don't expect to see a lot of detail in these galaxies. What you will be able to see is the differences in shapes between the group. Most are spirals and the remainder are what are known as irregulars and elliptical. How many of these galaxies you'll see will be determined by how long you let your eyes dark-adapt, the local weather conditions, and the overall brightness of the sky. Don't be disappointed if at first you do not see a hundred galaxies. Try using averted vision — looking through the corners of your eyes — or moving to a darker location for observing. Most of all, enjoy yourself as you peer into the far reaches of the universe and the distant past. This is Lawnchair Astronomy at its finest.

Libra:
Nothing but Space and Green Stars

Libra sits on the ecliptic about forty-five degrees up from the horizon. Like Corvus, which lies below Virgo, the constellation Libra is a diamond-shaped aggregation of stars that resides just to the right of Scorpius and the summer Milky Way.

Home to very few Lawnchair-accessible sights, it is most notable for its two primary stars, Alpha and Beta Libra, or the Southern and Northern Claws. A debate exists concerning the color of Beta Libra, with some reporting it to be green in color, which is usually only witnessed in the presence of a larger red companion star. Throughout the universe there are less than a handful of green stars, and those that have been cataloged have always been associated with a red giant partner. It is thought that the red light emitted from the companion may alter the true color of the smaller component, which is not really producing green light.

Beta Libra is the brightest star in the area between Libra and Scorpius. If you find that it is indeed green, you'll be in the minority of those observers that interpret this shade as such. If it is green, though, then it is the only single star in the known universe that can make that claim. See what you can detect when you get the chance to see it. Facing Libra, Beta will be the top star of the diamond. You won't want to stay in Libra too long, though, not with Scorpio nipping at your eyepiece!

Scorpio:
The Ruby Red Heart of Summer

Looking across the southern horizon on any dark summer evening, you will come to what looks to me more like an archer's bow and arrow than a scorpion. This is the constellation Scorpius. The brightest star here, the middle star of the archer's arrow, is the very well known ruby-red Antares.

The Chinese called it the Fire Star; the Greeks referred to it as the Rival of Mars; and both the French and Romans called it Le Coeur du Scorpion, the Heart of the Scorpion. Unmistakably red, you'll be impressed when you train even binoculars on this star. The only other like it is the red giant Betelguese in winter's Orion. These two candidates for the big one, a supernova, currently pulsate erratically though slowly, and are both energetic radio sources. Though if Antares were to go boom today, we'd

wouldn't know about it for five hundred years. Or it may have already gone supernova or turned into a black hole and we'll find out soon, tomorrow perhaps. Isn't this a great science?

Antares has a companion — several actually. But its closest of note is its secondary, a small *green* star. Seen only in larger instruments of 10 inches or more, this companion is a prime example of what we talked about earlier in the section on Libra. Green stars only exist accompanying large red stars. In a clear-air horizon, this companion should be visible as an emerald sparkle just alongside the large red giant.

Just a hop away, minus the skip and/or jump, mind you, is the easiest and largest globular cluster you can find all summer. Just point at Antares and shift over toward the west less than a degree and a half, and you're there. M-4 is one of those

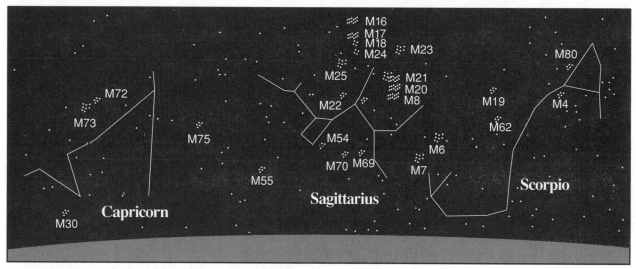

Summer's sting: Scorpio and the constellations of the heart of summer

jewels that you will go back to see every summer.

Elsewhere in the constellation of Scorpius, as you'll learn from a sweep of the area, are fantastic star clouds and groups of clusters, M-6 and M-7, to name a few. Both are easily seen in binoculars alone and are that much more impressive in small to medium-size telescopes. The entire southern horizon is filled with the fuzzies and sparkles of the summer Milky Way! It's also a great excuse to invite some friends over who can appreciate all the finer aspects of superior Lawnchair, fire up the old BBQ, and consume some liquid refreshments. I guarantee it will be a night you won't soon forget!

Sagittarius:
Tea Time in the Milky Way

The Milky Way in Sagittarius centers on the teapot, the major asterism associated with Sagittarius. While looking South by ten o'clock on any clear summer night, this grouping of stars is hard to miss. This teapot is not all there is to the constellation Sagittarius, though. Just as the Big Dipper is not all there is to Ursa Major, the teapot is only the top right corner of a larger constellation. The teapot section, though, is where all the action is, especially since it is completely immersed in the Milky Way.

With a pair of binoculars the area in and around the teapot is

M8: The Lagoon Nebula

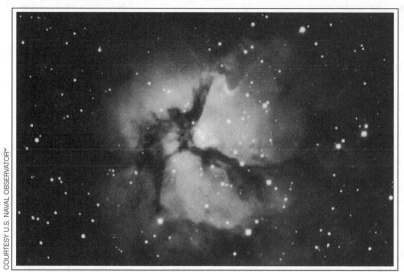

Major major: The Trifid Nebula (M-20)

you'll be amazed at how rich this portion of the sky really is.

Some of the most famous and photographed sights reside here. The Lagoon Nebula, M-8, was described by the author Robert Burnham, Jr., as: "one of the finest of the diffuse nebulae," and "plainly visible to the naked eye as a comet-like patch just off the mainstream of the Sagittarius Milky Way." Once you've found the teapot, M-8 is seemingly floating like a small puff of steam just above the spout section. This object is more than just a cloud of gas, though.

This mongrel of a deep-sky object is in itself worth a small library's worth of descriptive notes. Combined with elements of star clusters, dark lanes, and more traditional nebulae attributes, M-8 has always been an object of intense study and debate. The Lagoon Nebula gets its name from the dark rift that intersects its central region. More like a channel than a lagoon,

filled to the brim with bright stars, clusters, light and dark gas-dust clouds, as well as countless dim stars in seemingly endless fields of pinpoint sparkles. With medium- to large-size telescopes the teapot can simply be awe-inspiring. Of course, the darker your night and location is, the better. And once your eyes become dark-adapted this superimposed blotch changes from coal to gray-black.

Throughout the Lagoon Nebula are additional dark globules that some astronomers believe might perhaps be the precursors to star formation seen at the earliest stage. And then to add even more to an already full dance card, a loose cluster of bright stars floats among all this nebulosity. Throw in more than just a smattering of dim background stars to pepper the field and you've got a good idea of what to expect from M-8.

In the same field, using a low-power wide-angle eyepiece, you should also be able to see the Trifid Nebula, M-20, an object that is considered a "major major" in the world of astrophotography. With additional magnification and against a dark night, this sight will take quite a bit of staring at before you become bored with it. Nebulous areas separated by dark jets and lanes, the Trifid Nebula is unique in its appearance in that it seems to be symmetrically divided. In the same field with the Trifid is M-21, a small impact star cluster.

Speaking of clusters, M-22, lying just to the left of the teapot's dome star, is second in stature only to the great globular in Hercules. Estimated at nearly a half million members strong (500,000 stars), this bright cluster is one of the special jewels that will greet you as you tour Sagittarius.

M-22: An impressive cluster

The teapot is indeed a busy little area of sky. Since the earliest times, people have stared for hours at this small section of heaven. You should never feel alone as you find yourself staring. Sagittarius is at the heart of the galaxy, and of Lawnchair Astronomers everywhere.

Capricorn:
Last but Not Least, a Brief Encounter

Last, but not quite least on our zodiacal list, is Capricorn. This large arrowhead-shaped constellation points down toward the eastern horizon and arrives in the late evenings of July and August. If you can see down far enough, you'll see the small obscure constellation Microscopium, which when seen from the tropics at this time of the year is quite high in the sky.

Capricorn houses the star cluster called M-30. At magnitude 7.5, binoculars can detect it as a fuzzy patch. A medium-size telescope will easily resolve some of its outer members. Other than that, the constellation contains a handful of double stars and little else of interest to deep-sky fans.

Capricorn is where Jean Galle first discovered the planet Neptune. For Lawnchair Astronomers who have tasted the universe at large, Capricorn will be at most a brief encounter.

The zodiac is just the starting point for your adventures out into the universe. Using the constellations as road maps to some of the most incredible sights that humans can see, as a Lawnchair Astronomer you now can lead the way. Don't hesitate to show anyone who seems interested just what it is that makes you keep coming back for more. Spreading of the Astronomitus virus is considered both socially acceptable behavior and politically correct. Long live Lawnchair Astronomy!

LAST NOTES FROM THE LAWNCHAIR

Everyone will want to look at something different. That's a given. Some of your friends will want to see Saturn and its magnificent ring system, while others will want to see the Andromeda Galaxy for themselves. Your duty, as an accomplished Lawnchair Astronomer, is to assist them by gently pointing out some of the wonderful sights that can be found in a backyard sky. To aid you in your pursuit, here are some final words and valuable tips for attaining the coveted joy that is Lawnchair Nirvana.

Dark-Adapted Eyes — Natural Night Vision

The nature of Lawnchair Astronomy generally dictates that you be outside at night. That being the case, *seeing* is of greatest importance. Light, any light, other than that coming from the stars, is the last thing that you'll want to encounter.

Once your eyes dark-adapt, that is, once they get used to the lower light levels found during a typical backyard evening, they will be better suited to "seeing" the often dim and wispy objects waiting for you in the night sky. This process can take nearly twenty minutes and will take longer should you re-expose your eyes to any kind of lighting. In a previous section, The List, there is reference to a red-tinted flashlight for looking up information on a starmap. The reason is: Red-tinted light does not as adversely affect dark-adapted eyes, and therefore it is the only kind of lighting recommended to accompany these activities.

Light Pollution

While we're on the subject of light, light pollution is another factor that Lawnchair and professional astronomers often have to deal with. Brightened horizons in the direction of major cities are, to say the least, disturbing, and generally prevent one from seeing anything of interest in this already atmospherically turbulent area of the sky. For years, excessive city and town lighting conditions have been cause for Lawnchair Astronomers to head for darker, more rural skies. Lawnchairs are often seen strapped to the cartops of terminally stricken, though smiley-faced, astronomitus victims, along with telescopes and tripods, ice chests, and tents — for those extended multi-evening, "boy, sure is dark out here" type of stays. Enthusiasts who'd take this route are advised to travel at least fifty miles from the offending light source in order to escape this modern-day scourge. The farther, the better, is my advice.

Averted Vision

Another neat little trick is to use a technique called "averted vision" to see objects that lie just on the threshold of your eye's light-gathering ability. The center of your eye, well used to the daily bombardment of both natural and artificial light, is less sensitive to dim light than are the side portions of your eye. Rather than looking directly at a dim object (with or without a telescope or binoculars), try looking away to the right or left, and you just may glimpse a little more detail than you could have looking straight on.

Use a Cool Telescope

Perhaps more important in the fall, winter, and spring seasons, when the outside temperatures are often drastically different (colder) from those indoors (where the telescope lives), your instrument will perform better if it is "equalized," temperature-wise, to the environment in which you'll use it. When you decide on a time to go out and stargaze, bring your scope out first and let it cool a bit. Warm air inside the tube, especially in a larger telescope that has just been brought outdoors, can make seeing less perfect. For the same reasons that you wouldn't want to look at objects embedded in the atmosphere-thick horizon, objects seen through a tube that hasn't cooled sufficiently can seem distorted and seemingly unfocusable. Fortunately, this effect dissipates with time once the air temperature inside and outside the tube *equalizes*.

Reality in the Eyepiece

Finally, don't expect to see astrophoto images in a telescope. Those types of images are typically produced via extremely long-exposure processes that the eye is just plain incapable of. What you will see, however, often looks cartoonish or ghostlike. What the photographs won't produce is the feeling of depth that you'll encounter when viewing areas of the sky where the star fields are partially blocked out by dark nebula. Or when you zoom in on one of the hundreds of globular star clusters — their awesome brightness and sparkle just can't be captured on film.

Lastly...

The very last and perhaps most important piece of advice I can offer is to remind you again to dress warmly, on your feet especially. If you're at all like most Lawnchair Astronomers, you'll often find yourself standing out in the cold for hours on end, gushing with excitement over what's floating across your eyepiece. In order to comfortably enjoy the experience, make it a point to dress according to the weather, even if you have to don the NASA-grade snow boots.

I hope this book has done a little to prepare you for what is one of the most enjoyable hobbies that you can pursue today. Remember, this is not something you'll do all in one evening. Lawnchair Astronomy can provide you with a lifetime's worth of treasures, treasures that you can in turn share with your children, your friends and family, and anyone who knows the real value of a comfortable and well-equipped Lawnchair. I'll leave you now with one last thing to ponder — an old and well-worn Lawnchair saying, "What do you mean, you forgot the bug spray?"

Suggested Additional Materials and Reading

Sky & Telescope magazine

Astronomy magazine

Burnham's Celestial Handbook

Whitney's Star Finder

The Stargazer's Bible

Peterson's Field Guide to the Stars and Planets

Notes